Lecture Notes in Physics

The Lecture Notes in Physics

The series Lecture Notes in Physics (LNP), founded in 1969, reports new developments in physics research and teaching – quickly and informally, but with a high quality and the explicit aim to summarize and communicate current knowledge in an accessible way. Books published in this series are conceived as bridging material between advanced graduate textbooks and the forefront of research to serve the following purposes:

- to be a compact and modern up-to-date source of reference on a well-defined topic;
- to serve as an accessible introduction to the field to postgraduate students and nonspecialist researchers from related areas;
- to be a source of advanced teaching material for specialized seminars, courses and schools.

Both monographs and multi-author volumes will be considered for publication. Edited volumes should, however, consist of a very limited number of contributions only. Proceedings will not be considered for LNP.

Volumes published in LNP are disseminated both in print and in electronic formats, the electronic archive is available at springerlink.com. The series content is indexed, abstracted and referenced by many abstracting and information services, bibliographic networks, subscription agencies, library networks, and consortia.

Proposals should be sent to a member of the Editorial Board, or directly to the managing editor at Springer:

Dr. Christian Caron
Springer Heidelberg
Physics Editorial Department I
Tiergartenstrasse 17
69121 Heidelberg/Germany
christian.caron@springer.com

Heinrich Schwoerer   Joseph Magill
Burgard Beleites   (Eds.)

Lasers and Nuclei

Applications of Ultrahigh Intensity Lasers
in Nuclear Science

 Springer

Editors

Heinrich Schwoerer
Burgard Beleites
Friedrich-Schiller-Universität
Institut für Optik und Quantenelektronik
Max-Wien-Platz 1
07743 Jena, Germany
E-mail: beleites@ioq.uni-jena.de
schwoerer@ioq.uni-jena.de

Joseph Magill
European Commission
Joint Research Centre
Institute for Transuranium Elements
Hermann-von-Helmholtz-Platz 1
76344 Eggenstein- Leopoldshafen
Germany
E-mail: Joseph.Magill@cec.eu.int

H. Schwoerer et al., *Lasers and Nuclei*,
Lect. Notes Phys. 694 (Springer, Berlin Heidelberg 2006), DOI 10.1007/b11559214

Library of Congress Control Number: 2006921739

ISSN 0075-8450
ISBN-10 3-540-30271-9 Springer Berlin Heidelberg New York
ISBN-13 978-3-540-30271-1 Springer Berlin Heidelberg New York

Springer is a part of Springer Science+Business Media
springer.com
© Springer-Verlag Berlin Heidelberg and European Communities 2006
Printed in The Netherlands

The use of general descriptive names, registered names, trademarks, etc. in this publication does not imply, even in the absence of a specific statement, that such names are exempt from the relevant protective laws and regulations and therefore free for general use.

Legal notice/disclaimer:
Neither the European Commission nor any person acting on behalf of the Commission is responsible for the use which might be made of the information contained in this document.

Typesetting: by the authors and techbooks using a Springer LATEX macro package
Cover design: *design & production* GmbH, Heidelberg

Printed on acid-free paper   SPIN: 11559214   57/techbooks   5 4 3 2 1 0

Preface

The subject of this book is the new field of laser-induced nuclear physics. This field emerged within the last few years, when in high-intensity laser plasma physics experiments photon and particle energies were generated, which are high enough to induce elementary nuclear reactions. First successful nuclear experiments with laser-produced radiation as photo-induced neutron disintegration or fission were achieved in the late nineties with huge laser fusion installations like the VULCAN laser at the Rutherford Appleton Laboratory in the United Kingdom or the NOVA laser at the Lawrence Livermore National Laboratory in the United States. But not before the same physics could be demonstrated with small tabletop lasers, systematic investigations of laser-based nuclear experiments could be pushed forward. These small laser systems produce the same light intensity as the fusion laser installations at lower laser pulse energy but much higher shot repetition rates. Within a short and lively period all elementary reactions from fission, neutron and proton disintegration, and fusion to even cross section determinations were demonstrated.

From the very beginning a second focus beyond proof of principal experiments was laid on the investigation of the unique properties of high-energy laser plasma emission in the view of nuclear physics topics. These special features are manifold: the ultrashort duration of all photon and particle emissions in the order of picoseconds and shorter, the very small source size due to the small interaction volume of the laser light with the target matter and, not to underestimate, the high flexibility and compactness of the radiation source installation compared to conventional accelerator- or reactor-based installations.

With these novel experimental possibilities, a variety of potential applications in science and technology comes into mind. Most obvious is the diagnostic and characterization of the relativistic laser plasma with the help of nuclear activation, which is the only method available to detect ultrashort pulses of high-energy radiation and particles. A second range of potential applications is the transmutation of nuclei. Because of the diversity of projectiles, generated or accelerated in the laser plasma, all reaction paths with photons, protons, ions, and neutrons are accessible. Realistic ideas cover the production of radioisotopes for medical purposes as well as for the investigation of transmutation scenarios for long-lived radioactive nuclei for the nuclear fuel cycle. Finally, the extreme energy density in the laser plasma in combination with the large flux of high-energy particles offers also new possibilities for fundamental nuclear science like the study of astrophysical problems in the laboratory.

The scope of the book, as well as of the international workshop "Lasers & Nuclei" held in Karlsruhe in September 2004, which stimulated the book, is to bring together, for the first time, laser and nuclear scientists in order to present the current status of their fields and open their minds for the experimental and theoretical potentials, needs, and constraints of the new interdisciplinary work. The book starts with an introduction to the theoretical background of laser–matter interaction and overview reports on the state of research and technology. In the second part, detailed reports on the state of research in laser acceleration of particles and laser nuclear physics are given by leading scientists of the field. The third part discusses potential applications of these new joint activities reaching from laser-based production of isotopes, the physics of nuclear reactors through neutron imaging techniques all the way to fundamental physics in nuclear astrophysics and pure nuclear physics.

With its broad and interdisciplinary spectrum the book shall stimulate thinking beyond the traditional paths and open the mind for the new activities between laser and nuclear physics.

Jena and Karlsruhe   *Heinrich Schwoerer*
February 2006   *Joseph Magill*
                                                             Burgard Beleites

Contents

List of Contributors

Didier Besnard
CEA/DAM Île de France, BP12
91680 Bruyères le Châtel
France
didier.besnard@cea.fr

Charles Bowman
ADNA Corporation
1045 Los Pueblos
Los Alamos
NM 87544, USA
Cbowman@cybermesa.com

Friederike Ewald
Institut für Optik
und Quantenelektronik
Friedrich-Schiller-Universität
Max-Wien-Platz 1
07743 Jena, Germany

Fazia Hannachi
Centre d'Etudes Nucléaires de
Bordeaux Gradignan
Université Bordeaux 1
le Haut Vigneau
33175 Gradignan cedex
France
hannachi@cenbg.in2p3.fr

Joachim Hein
Institut für Optik und
Quantenelektronik
Friedrich-Schiller-Universität
Max-Wien-Platz 1
07743 Jena
Germany
jhein@ioq.uni-jena.de

Kazuo Imasaki
Institute for Laser Technology
2-6, Yamada-Oka Suita
Osaka
565-0871 Japan
kzoimsk@ile.osaka-u.ac.jp

Eberhard Lehmann
Spallation Neutron Source Division
(ASQ)
Paul Scherrer Institut
CH-5232 Villigen PSI
Switzerland
eberhard.lehmann@psi.ch

Joseph Magill
European Commission
Joint Research Centre
Institute for Transuranium Elements
Postfach 2340
76125 Karlsruhe
Germany
Joseph.Magill@cec.eu.int

Victor Malka
Laboratoire d'Optique Appliquèe
ENSTA –
CNRS UMR 7639
Ecole Polytechnique
Chemin de la Hunière
91761 Palaiseau cedex
France
victor.malka@ensta.fr

Paul McKenna
Department of Physics
University of Strathclyde
Glasgow, G4 0NG
Scotland, UK
p.mckenna@phys.strath.ac.uk

Charles Rhodes
Laboratory for X-Ray Microimaging
and Bioinformatics
Department of Physics
University of Illinois at Chicago
Chicago, IL 60607-7059
USA
rhodes@uic.edu

Lynne Robson
Department of Physics
University of Strathclyde
Glasgow, G4 0NG
Scotland, UK
l.robson@phys.strath.ac.uk

Heinrich Schwoerer
Institut für Optik und
Quantenelektronik
Friedrich-Schiller-Universität
Max-Wien-Platz 1
07743 Jena
Germany
schwoerer@ioq.uni-jena.de

Toshiyuki Shizuma
Advanced Photon
Research Center
Kansai Research Establishment
Japan Atomic Energy Research
Institute
8-2 Umemidai, Kizu
619-0215 Kyoto
Japan
shizuma@popsvr.tokai.jaeri.go.jp

Tomaž Žagar
Jožef Stefan Institute
Jamova 39
1000 Ljubljana
Slovenia
tomaz.zagar@ijs.si

Part I

Fundamentals and Equipment

1

The Nuclear Era of Laser Interactions: New Milestones in the History of Power Compression

A.B. Borisov[1], X. Song[1], P. Zhang[1], Y. Dai[2], K. Boyer[1], and C.K. Rhodes[1,2,3,4]

[1] Laboratory for X-Ray Microimaging and Bioinformatics, Department of Physics, University of Illinois at Chicago, Chicago, IL 60607-7059 USA
[2] Department of Bioengineering, University of Illinois at Chicago, Chicago, IL 60607-7052 USA
[3] Department of Computer Science, University of Illinois at Chicago, Chicago, IL 60607-7042 USA
[4] Department of Electrical and Computer Engineering, University of Illinois at Chicago, Chicago, IL 60607-7053, USA
rhodes@uic.edu

Abstract. A brief review of the history of power compression over a range encompassing approximately 40 orders of magnitude places laser–nuclear interactions roughly at the logarithmic midpoint of the scale at approximately $10^{20}\,\mathrm{W/cm^3}$. The historical picture also motivates four conclusions, specifically, that (1) foreseen developments in power compression will enable laser-induced coupling to all nuclei, (2) conventional physical mechanisms will encounter a limit of $\Omega_\alpha \sim 10^{30} - 10^{31}\,\mathrm{W/cm^3}$, a value approximately 10^{10} above the presently demonstrated capability, (3) the key to reaching the Ω_α limit is the generation of relativistic/charge-displacement self-trapped channels with multikilovolt X-rays in high-Z solids, a concept named "photon staging," and (4) penetration into the $10^{30} - 10^{40}\,\mathrm{W/cm^3}$ zone, the highest range known and the region represented by processes of elementary particle decay, will require an understanding of new physical processes that are presumably tied to phenomena at the Planck scale.

1.1 History of Power Compression

The goal of achieving the coupling of laser radiation to nuclear systems has an extensive history, one that spans a range of approximately three decades. An excellent source of information on this history is the comprehensive landmark article [1] by Baldwin, Solem, and Gol'danskii entitled "Approaches to the Development of Gamma-Ray Lasers." In light of the progress made during the 25 years since the publication of this important piece, it is now possible to foresee

A.B. Borisov et al.: *The Nuclear Era of Laser Interactions: New Milestones in the History of Power Compression*, Lect. Notes Phys. **694**, 3–6 (2006)
www.springerlink.com © Springer-Verlag Berlin Heidelberg and European Communities 2006

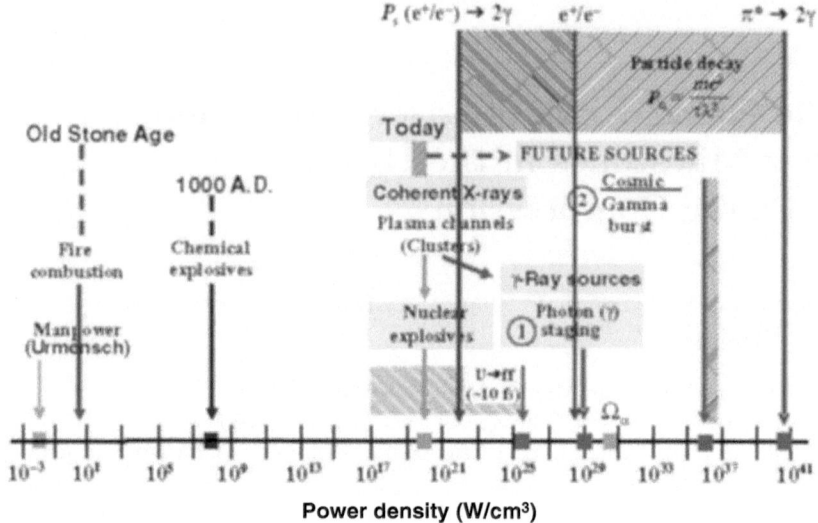

Fig. 1.1. History of power technology that illustrates a range spanning more than 10^{40} in power density. This range corresponds physically to the development from raw manpower to rapid particle decay. The present status, roughly situated at 10^{20} W/cm^3, is experimentally associated with nuclear explosives, laser-induced nuclear fission [2, 3, 4], and coherent X-ray amplification [5, 6]. An estimated, power density limit that could be achieved by the channeling of multikilovolt X-rays in a high-Z solid, designated as $\Omega_\alpha \sim 10^{30} - 10^{31}$ W/cm^3, is indicated

the practical production of the power densities sufficient for amplification in the γ-ray region associated with nuclear transitions.

The history of power compression that is presented in Fig. 1.1 illustrates the presence of several developmental epochs. Each is separated by a factor of approximately 10^{10} and each stage marks a technological breakthrough. Also apparent from this history is the fact that the attainment of each new level in power density generally manifests itself in two forms. Initially, a state of matter is produced from which a largely uncontrolled energy release is obtained, such as that associated with a chemical explosive. This signal event is subsequently followed by an additional innovation, in this case of conventional explosives the cannon, that generates an ordered controlled outcome that channels the energy. Control is thus conjoined with power at each stage of the development. At the level of 10^{20} W/cm^3, as shown in Fig. 1.1, nuclear explosives and coherent X-ray amplification, respectively, correspond to the uncontrolled and controlled forms. In this case, the innovation leading to the multikilovolt X-ray amplification is the combination of (1) a new concept for amplification, which involves the creation of a highly ordered composite state of matter incorporating ionic, plasma, and coherent radiative components, with (2) the use of two recently discovered ($\sim$1990) forms of radially symmetric energetic matter, namely, hollow atoms and self-trapped plasma channels. The present status of power compression places us roughly at the logarithmic midpoint ($\sim 10^{20}$ W/cm^3) of the power density scale. Overall, this level is experimentally represented by three phenomena, (1) nuclear explosives, (2) laser-induced fission [2, 3, 4], whose limiting value of approximately 10^{25} W/cm^3 for the complete fission of solid uranium in a time of approximately 10 fs is indicated, and (3) X-ray amplification on Xe(L) hollow atom transitions [5, 6] at $\lambda \sim 2.8$ Å. Extension of the power density to significantly higher values ($\sim 10^{30}$ W/cm^3) is projected with the achievement of channeled propagation of multikilovolt X-rays in a high-Z solid, a process called "photon staging."

1.2 Conclusions

The information available at the time of this gathering in Karlsruhe is sufficient to hazard four conclusions. They are as follows: (1) predicted advances in power compression will enable laser-induced coupling to all nuclei, (2) a power density limit $\Omega_\alpha \sim 10^{30} - 10^{31}$ W/cm^3 can be reached with the use of conventional presently understood physical processes, (3) the key to reaching the Ω_α limit is the production of relativistic/charge-displacement self-trapped channels with multikilovolt X-rays in high-Z solids, and (4) penetration into the approximately $10^{30} - 10^{40}$ W/cm^3 zone will require an understanding of fundamentally new physics, most probably tied to the Planck scale. For the latter, a rich observational basis exists and a conceptual synthesis has been hypothesized [7, 8, 9], but a full theoretical picture remains undeveloped.

Acknowledgments

This work was supported in part by contracts with the Office of Naval Research (N00173-03-1-6015), the Army Research Office (DAAD19-00-1-0486 and DAAD19-03-1-0189), and Sandia National Laboratories (1629, 17733, 11141, and 25205). Sandia is a multiprogram laboratory operated by the Sandia Corporation, a Lockheed Martin Company for the United States Department of Energy, under contract no. DE-AC04-94AL85000.

References

1. G.C. Baldwin, J.C. Solem, V.I. Gol'danskii: Rev. Mod. Phys. **53**, 687 (1981)
2. K. Boyer, T.S. Luk, C.K. Rhodes: Phys. Rev. Lett. **60**, 557 (1988)
3. K.W.D. Ledingham, I. Spencer, T. McCanny, R. Singhal, M. Santala, E. Clark, I. Watts, F. Beg, M. Zepf, K. Krushelnick, M. Tatarakis, A. Dangor, P. Norreys, R. Allott, D. Neely, R. Clark, A. Machacek, J. Wark, A. Cresswell, D. Sanderson, J. Magill: Phys. Rev. Lett. **84**, 899 (2000)

4. T.E. Cowan, A. Hunt, T. Phillips, S. Wilks, M. Perry, C. Brown, W. Fountain, S. Hatchett, J. Johnson, M. Key, T. Parnell, D. Pennington, R. Snavely, Y. Takahashi: Phys. Rev. Lett. **84**, 903 (2000)
5. A.B. Borisov, X. Song, F. Frigeni, Y. Koshman, Y. Dai, K. Boyer, C.K. Rhodes: J. Phys. B: At. Mol. Opt. Phys. **36**, 3433 (2003)
6. A.B. Borisov, J. Davis, X. Song, Y. Koshman, Y. Dai, K. Boyer, C.K. Rhodes: J. Phys. B: At. Mol. Opt. Phys. **36**, L285 (2003)
7. Y. Dai, A.B. Borisov, J.W. Longworth, K. Boyer, C.K. Rhodes: In: *Proc. Int. Conf. Electromagnet. Adv. Appl.*, Politecnico di Torino, Torino, Italy, 1999, ed. R. Graglia, p. 3
8. Y. Dai, A.B. Borisov, J.W. Longworth, K. Boyer, C.K. Rhodes: Int. J. Mod. Phys. A **18**, 4257 (2003)
9. Y. Dai, A.B. Borisov, J.W. Longworth, K. Boyer, C.K. Rhodes: Adv. Stud. Contemp. Math. **10**, 149 (2005)

2

High-Intensity Laser–Matter Interaction

H. Schwoerer

Institut für Optik und Quantenelektronik, Friedrich-Schiller-Universität,
Max-Wien-Platz 1, 07743 Jena, Germany
schwoerer@ioq.uni-jena.de

2.1 Lasers Meet Nuclei

A visible laser beam can be used to set neutrons free, to induce the fusion between nuclei, or even to fission a nucleus. The energy of one laser photon is about 1 eV, whereas the energy required to fission a uranium nucleus amounts to 10 million electronvolts. How can that be?

The trick is that the intensity of the laser light field of today's most powerful lasers is so high that the interaction of the light with matter is completely dominated by the electromagnetic field rather than by single photons. Or, in other words, the interaction physics has left the regime of classical nonlinear optics and has emerged into the new domain of relativistic optics.

We will see that this situation has manifold consequences. Relativistic optics or relativistic interaction between light and matter, even though this phrase is literately not quite correct, starts when the quiver energy of an electron in the light field approaches the energy of its rest mass divided by the square of the speed of light. This occurs at a light intensity of $2 \times 10^{18}\,\mathrm{W/cm^2}$ (at the wavelength of today's intense lasers of 800 nm). Today's ultrashort pulse, high-intensity lasers are capable of generating intensities two orders of magnitude higher than this value. Therefore, experimental laser physics has truly entered this novel regime.

The situation in the focus of such a laser can be demonstrated with a simple analogy (see Fig. 2.1): If one focusses all sunlight incident on the earth with a big enough lens onto the tip of a pencil ($0.1\,\mathrm{cm^{-2}}$), the intensity in that spot would be $10^{20}\,\mathrm{W/cm^2}$. In the focus of that laser pulse, the electric field strength is more than $10^{11}\,\mathrm{V/cm}$, a value almost hundred times higher than the field binding the electron and the proton in the hydrogen atom. The light pressure onto a solid in the laser focus reaches several Gbar. Through directed acceleration of electrons, currents of many TeraA/cm² and magnetic fields of several thousand Tesla are generated. And, finally a macroscopic amount of dense matter is heated to millions of degrees. These states of matter and fields of that size exist in stars, in the vicinity of black holes, and in galactic jets.

H. Schwoerer: *High-Intensity Laser–Matter Interaction*, Lect. Notes Phys. **694**, 7–23 (2006)
www.springerlink.com   © Springer-Verlag Berlin Heidelberg and European Communities 2006

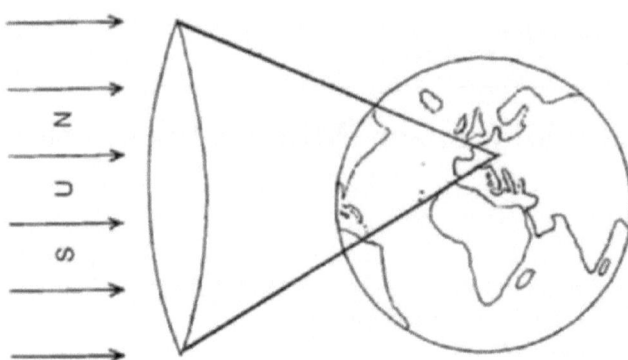

Fig. 2.1. If the focal spot size of the sunlight lens is $0.1\,\text{mm}^2$, the intensity there would be $10^{20}\,\text{W/cm}^2$. This value can be reached with state-of-the-art high-intensity laser systems, admittedly, only for $10^{-12}\,\text{s}$

In the laboratory however, they can be produced only in a controlled way in high-intensity laser plasmas.

Before we describe the fundamental interaction between such intense light fields with matter, we quickly have to characterize the light fields themselves in the following Sect. 2.2. In Sect. 2.3, we will introduce the basic mechanisms of laser–matter interaction at relativistic intensities, starting from the free electron in a strong electromagnetic wave all the way to the forced wakefield or bubble acceleration. Section 2.4 covers the generation of Bremsstrahlung in the multi-MeV range, and the final Sect. 2.5 describes the fundamentals of proton and ion acceleration with intense laser pulses.

2.2 The Most Intense Light Fields

The intensity of a laser pulse is given by the pulse energy E divided by the pulse duration τ and the size of the focal area A. In order to reach relativistic intensities, E has to be large whereas τ and A must be as small as possible. For technical and financial reasons basically two combinations of these parameters exist in real lasers: a high-energy version and an ultrashort version. The high-energy laser systems typically deliver pulse energies of hundreds to thousand Joules within pretty short pulses below 1 ps. The ultrashort laser systems concentrate their pulse energy in the range of 1 J within much less than 100 fs. Because of better focusability in the latter case, both types are able to generate the same maximum intensity. Another difference between the two types of lasers is the rate of shots. Thermal effects within the laser typically limit the high-energy systems to one shot within half an hour whereas the ultrashort systems can operate around 10 Hz. Because of the high investments and operational costs of a high-energy laser system, these are run by national laboratories like the Rutherford Appleton Laboratory in the United

Kingdom (VULCAN PetaWatt laser [1], which is currently the strongest laser system in the world) and are typically used for proof of principal experiments. Ultrashort high-intensity lasers, on the other hand, can be operated by smaller institutions like, for example, the Laboratoire d'Optique Appliquée in France [2] or the JETI laser at the Jena University and are used for more systematic investigations.

In order to understand the primary interaction of an intense laser pulse with matter, we have to describe the temporal and spatial structure of typical real laser pulses (see Fig. 2.2): Almost independent on the type of the high-power laser, the laser pulses usually consist of three contributions within different temporal regimes.

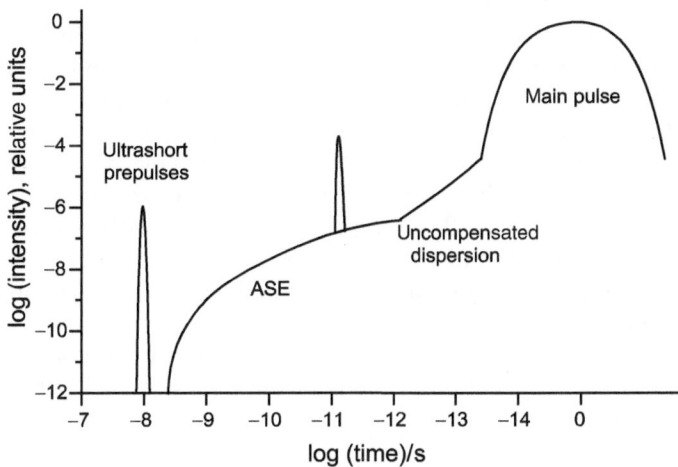

Fig. 2.2. Schematic representation of the temporal structure of an intense laser pulse. The ultrashort main pulse is surrounded by a picosecond-scale region of uncompensated dispersion, spontaneous amplified emission on a nanosecond scale, and possibly small prepulses as short as the main pulse. The relative intensities of all these contributions as shown in the drawing are more or less typical values, but in reality vary with the quality of the specific laser system

First, the central ultrashort main pulse, which usually exhibits a Gaussian temporal shape of $\tau = 30\,\text{fs}$ to 1 ps full width at half maximum. The power $P = E/\tau$ in the central peak is used as a characteristic measure of the laser system and ranges from tens of TW up to 1 PW (10^{12} to 10^{15} W) at present. However, the property which determines the interaction of light with matter and finally the energy range of particles and photons emitted from the interaction region is the power density $I = P/A$, where A is the illuminated area. The power density, or intensity, of nowadays strongest lasers reaches values of 10^{20} to $10^{21}\,\text{W/cm}^2$.

The second contribution to the temporal shape of the laser pulse in the focus is noncompensated angular and temporal dispersion. The shorter the laser pulses are, the broader is their spectral width and therefore the exact compensation of all dispersion in space and time picked up by the pulse within the laser system becomes more and more difficult. The uncompensated dispersion of a sub 100 fs pulse typically reaches out to about 500 fs to 1 ps and reaches a level of 10^{-4} to 10^{-3} of the maximum intensity. That is more than enough to ionize any matter. But, because of the short duration until the main laser pulse arrives, this plasma cannot evolve much and does not expand far into the vacuum. Therefore, the interaction with the main pulse is only slightly altered by the uncompensated dispersion.

However, the third contribution is of high importance for the fundamental interaction mechanisms: Because of amplified spontaneous emission (ASE) in the laser amplifiers, a long background surrounds the main laser pulse. It starts several nanoseconds in advance of the main pulse and reaches relative levels of 10^{-6} to 10^{-9} of the main pulse intensity, depending on the quality of the pulse-cleaning technology implemented in the laser system. Since the ionization threshold lies in the vicinity of 10^{12} W/cm^2, the prepulse due to ASE is sufficient to produce a preplasma in front of the target prior to the arrival of the main pulse. The preplasma expands with a typical thermal velocity of about 1000 m/s into the vacuum. Therefore, the underdense plasma can reach out tens or even hundreds of micrometers when the main laser pulse impinges on it.

This extended preplasma has two effects on the interaction. The first is that it affects the propagation of the light due to its density-dependent index of refraction n_{pl} (see, e.g., [3]),

$$n_{\mathrm{pl}} = \sqrt{1 - \frac{n_{\mathrm{e}}\mathrm{e}^2}{\gamma m_0 \epsilon_0 \omega_{\mathrm{L}}^2}} \, , \tag{2.1}$$

where n_{e} is the local free electron density, e the elementary charge, γ the Lorentz factor, m_0 the electron's rest mass, ϵ_0 the dielectric constant, and ω_{L} the laser frequency. From (2.1) follows that light can propagate only in a plasma with a density n_{e} smaller than the critical density n_{cr}:

$$n_{\mathrm{e}} < n_{\mathrm{cr}} = \gamma m_0 \epsilon_0 \omega_{\mathrm{L}} / \mathrm{e}^2 \, . \tag{2.2}$$

In the opposite case, $n_{\mathrm{e}} > n_{\mathrm{cr}}$, the light cannot propagate. For a laser wavelength of 800 nm, which is the center wavelength of all tabletop high-intensity lasers, the critical plasma density is $n_{\mathrm{cr}}(800 \text{ nm}) = 1.7 \cdot 10^{21}$ cm^{-3}, which is about three orders below solid state density. Consequently, laser light penetrating into the preplasma is absorbed and reflected if its density reaches the critical density. Furthermore, we will see below that the laser pulse, as it propagates through the underdense plasma, can modify the spatial distribution of the plasma density in such a way that the pulse is focussed or defocused by the plasma that has previously been generated by its leading wing.

The second effect of the extended plasma is that it makes available a long interaction length between the ultrashort light pulse and free electrons. In a long preplasma and even more pronounced in an ionized gaseous target, electrons can be trapped into the laser pulse or even into a fast-moving plasma wave. They are accelerated to relativistic energies over long distances, much longer than the spatial length of the laser pulse. We will describe the mechanisms of these acceleration processes in detail in Sect. 2.3.3.

Summing up the said and going back to Fig. 2.2, we realize that the fundamental interaction between ultra-intense light fields and matter requires an understanding of the mechanisms of electron acceleration in intense light fields and the understanding of the influence of the laser-generated preplasma on the propagation of the laser beam and on the acceleration of electrons.

2.3 Electron Acceleration by Light

2.3.1 Free Electron in a Strong Plane Wave

Let us start considering a weak, pulsed, linearly polarized, plane electromagnetic wave of frequency ω propagating in z-direction (see Fig. 2.3):

$$\boldsymbol{E}(z,t) = E_0 \cdot \hat{x} \cos(\omega t - kz). \tag{2.3}$$

The temporal duration of the pulse shall be long compared to the oscillation period of the light $2\pi/\omega$, and the electric field shall not depend on the transverse coordinates x and y. A free electron, originally at rest, oscillates in that

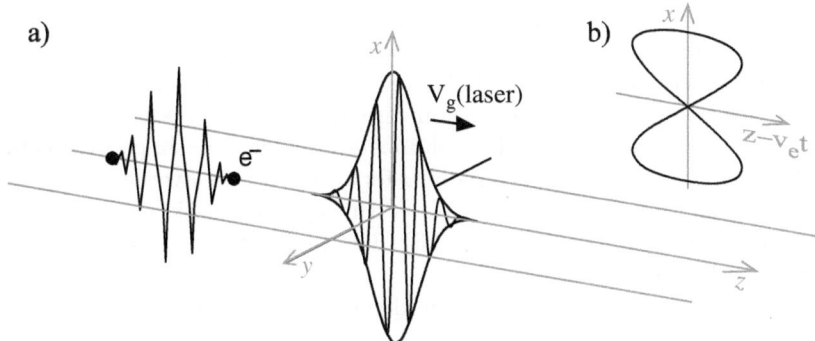

Fig. 2.3. (a) A relativistic laser pulse, propagating from *left* to *right* on the z-axis, has passed an electron. Its electric field points along the $\hat{x}$-direction, the magnetic field (not shown) into $\hat{y}$. The electron has moved along a zig-zag–shaped trajectory in the $\hat{x}$–$\hat{z}$-plane and stopped at rest after the passage. **(b)** "Figure of 8" electron trajectory in a frame moving with the mean forward velocity of the electron (averaged rest frame of the electron)

field along the direction of the E-field with a velocity v_x and a mean kinetic energy, also called quiver energy U_{osc},

$$v_x = \frac{eE}{m_0\omega} \sin(\omega t) \quad \text{and} \quad U_{osc} = \frac{e^2 E_0^2}{4m_0\omega^2}. \tag{2.4}$$

If the light pulse has passed through the electron, the electron is again at rest at the original position, no energy is transferred between light and electron. When we increase the electric field strength, finally the quiver velocity of the electron approaches the speed of light c and the quiver energy gets in the range of the rest energy $m_0 c^2$ of the electron and higher. In that regime, the magnetic field of the light wave cannot be neglected anymore in the interaction with the electron. The equation of motion of the electron has to be solved with the full Lorentz force

$$\boldsymbol{F}_{\mathrm{L}} = \frac{d\boldsymbol{p}}{dt} = -\mathrm{e} \cdot (\boldsymbol{E} + \boldsymbol{v} \times \boldsymbol{B}). \tag{2.5}$$

Since the velocity $\boldsymbol{v}$ due to the electric field is along $\hat{x}$ and the magnetic field $\boldsymbol{B}$ directs into $\hat{y}$ (see Fig. 2.3), the $\boldsymbol{v} \times \boldsymbol{B}$-term introduces an electron motion in $\hat{z}$-direction. Solving the relativistic equation of motion of the electron in the plane electromagnetic wave results in the momenta

$$p_x = -\frac{eE_0}{\omega m_0 c} \sin(\omega t - kz) = -a_0 \sin(\omega t - kz),$$

$$p_z = \left(\frac{eE_0}{\omega m_0 c}\right)^2 \sin^2(\omega t - kz) = \frac{a_0^2}{2} \sin^2(\omega t - kz). \tag{2.6}$$

Here we have introduced the relativistic parameter $a_0 = eE_0/\omega m_0 c$, which is the ratio between the classical momentum eE_0/ω as it results from (2.4) and the rest momentum $m_0 c$. We see from (2.6) that the electron oscillates in transverse direction $\hat{x}$ with the light frequency ω. The longitudinal velocity is always positive (in laser propagation direction) and oscillates with twice the light frequency. Overall, the electron moves on a zig-zag–shaped trajectory as displayed in Fig. 2.3. In a frame moving forward with the averaged longitudinal electron velocity $\langle \boldsymbol{v}_z \rangle_t = (eE_0/2c\omega)^2$, the electron undergoes a trajectory resembling an 8 (see inset in Fig. 2.3). The higher the relativistic parameter a_0, the thicker the eight.

The energy of the electron is can also be expressed with help of a_0:

$$E = \gamma m_0 c^2 = \left(1 + \frac{a_0^2 \sin^2(\omega t - kz)}{2}\right) \cdot m_0 c^2. \tag{2.7}$$

As we see from (2.6), the longitudinal momentum scales with the square of the intensity, whereas the transverse scales only linearly with I. Therefore, at high electric field strength the forward motion of the electron becomes dominant over the transverse oscillation.

Now we eventually have to come up with real numbers of electron energies versus electric field and light intensity: From (2.7) follows that the kinetic energy of the electron reaches its rest energy at $a_0 = \sqrt{2}$. For a laser wavelength of $\lambda = 800\,\mathrm{nm}$, this corresponds to an electric field strength of $E_0 \approx 5 \cdot 10^{12}\,\mathrm{V/m}$ and a light intensity of $I \approx 4 \cdot 10^{18}\,\mathrm{W/cm^2}$, following from $I = \frac{1}{2} c \epsilon_0 E_0^2$. At the currently almost maximum intensity of $10^{20}\,\mathrm{W/cm^2}$ the electric field is $E_0 \approx 3 \cdot 10^{11}\,\mathrm{V/cm}$ and the mean kinetic energy of the electron amounts to $6\,\mathrm{MeV}$. This relativistic electron energy gave rise to the term relativistic optics. Nevertheless, because of the planeness and transverse infinity of the light wave, our electron is again at rest after the pulse has passed it. It was moved forward, but no irreversable energy transfer took place.

2.3.2 An Electron in the Laser Beam, the Ponderomotive Force

An irreversable acceleration of an electron in a light field can be achieved only by breaking the transverse symmetry of the light field. This is obtained in real propagating light fields or laser beams, which all exhibit a transverse spatial intensity profile (see Fig. 2.4). In addition, we assume that the duration of the laser pulse is much longer than the oscillation period of the electromagnetic wave. Under these conditions the solution of the same equations of motion as above results in a force along the gradient of the intensity (see, e.g., [3]). This force is called the ponderomotive force F_pond, and it directs to lower intensities – the laser beam is a potential hill for the electron:

$$F_\mathrm{pond} = -\frac{e^2}{2m_0\omega^2} \cdot \nabla(E)^2 \,. \tag{2.8}$$

Again, there is no new mechanism involved in the ponderomotive potential. A simple way of looking at the ponderomotive force is the following: an electron close to the optical axis oscillates in the E-field direction while being pushed

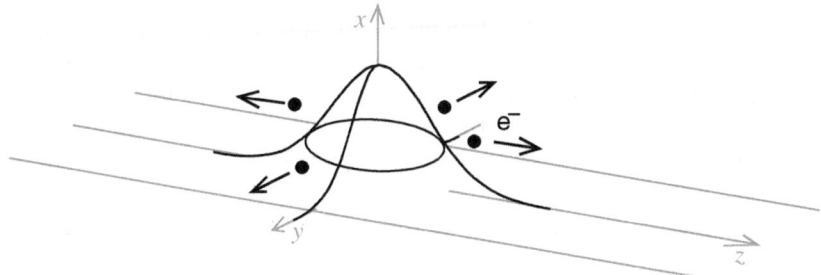

Fig. 2.4. The ponderomotive force F_pond: The drawing shows the spatial and temporal envelope of the electric field of a short laser beam, propagating in $\hat{z}$-direction. F_pond of the laser pulse acts onto electrons along the intensity gradients of the envelope field. Electrons are pushed out of the center of the beam, in radial as well as in forward and backward directions

forward because of the magnetic field. If its transverse travel amplitude comes to the order of the characteristic length of the spatial envelope $E/(dE/dx)$, the electron perceives at its outer turning point a smaller restoring force than close to the optical axis. It cannot come back to the original starting point. This is repeated during every oscillation until the electron finally leaves the beam with a residual velocity. As a numerical example, let us consider an electron initially on the optical axis of an $I = 10^{20}$ W/cm^2 laser beam. Once it has run down the ponderomotive potential, it has acquired an energy of 7.5 MeV; hence, it is relativistic.

However, since the initial position of the electron is equally distributed over the beam area and its initial momentum obeys a broad distribution; hence, at low energies, the final energy spectrum is again broad. In real experiments the details of the electron energy spectra strongly depend on a variety of experimental parameters like predominantly the extension and density distribution of the plasma, or the angle of incidence on a solid target and, of course, the intensity of the light field. Nevertheless, in most experiments the electron spectra follow an exponential distribution and a phenomenological temperature, the so-called hot electron temperature T_e can be attributed to them. Because of the variety of different experimental conditions, we will not try to give generalized scaling laws between the electron temperature and the laser intensity I, but restrict ourselves in this article to a few special cases. The first case relies on an early publication of Wilks et al. [4], which predicts a square root scaling of T_e versus I for the case of an ultrashort laser pulse, normally incident on a solid state target:

$$k_B T_e \simeq 0.511 \,\text{MeV}[(1 + I\lambda^2/1.37 \times 10^{18} \,\text{W/cm}^2\mu\text{m}^2)^{1/2} - 1], \qquad (2.9)$$

where k_B is the Boltzmann constant and λ the laser wavelength in micrometer.

Even though this correlation is derived under stringent conditions, interestingly it was verified in many experiments on very different laser systems (see, e.g., [5]). However, this relation describes the interaction of the laser pulse with overdense plasmas exhibiting a limited range of underdense prestructure.

The situation completely changes in extended underdense plasmas, for example, if the laser pulse is focussed into a gas jet. Here, collective processes of the plasma like resonant excitations can completely dominate the interaction and the electron acceleration. In the next section we will discuss the basic plasma electron acceleration mechanism, which is the wakefield acceleration.

2.3.3 Acceleration in Plasma Oscillations: The Wakefield

We will consider an extended and underdense plasma, which may be generated in a supersonic gas jet by the temporal prestructure of the main laser pulse. A laser pulse propagates through the plasma with a group velocity $v_g = c\,(1 - \omega_p^2/\omega^2)^{1/2}$, where $\omega_p = (n_e e^2/\epsilon_0 \gamma m_0)^{1/2}$ is the plasma frequency at the electron density n_e and ω the laser frequency. Our assumption of an

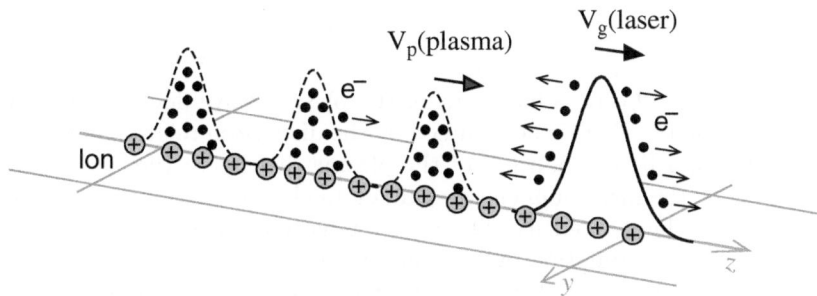

Fig. 2.5. Generation of a plasma wakefield and acceleration of electrons in the wakefield: The intense and ultrashort laser pulse (*solid envelope*) creates charge separation in the plasma, which can build up to a density wave (*dashed envelope*), travelling with a phase velocity close to the speed of light behind the laser pulse. Electrons entering this wakefield can be accelerated to relativistic energies

underdense plasma (plasma density n_e being smaller than the critical density $n_{cr} = \omega^2 \epsilon_0 \gamma m_0/e^2$) is equivalent to the condition that the laser frequency ω is larger than the plasma frequency ω_p and the light can propagate in the plasma.

As the laser pulse travels through the plasma, the ponderomotive force F_{pond} pushes the electrons out of its way (see Figs. 2.4 and 2.5). The z-axis component of F_{pond} acts on the electrons twice. First they are pushed forward by the leading edge of the pulse, and once they were surpassed by the pulse, they get a kick backward. This continuously introduces a longitudinal charge separation, which under certain laser and plasma conditions even can be driven resonantly (see, e.g., [6, 7, 8]). The simplest and most efficient case is given if the longitudinal length of the laser pulse is just half of the plasma wavelength $\lambda_p = 2\pi c/\omega_p$. The charge separation builds up to a charge density wave, whose phase velocity v_p is approximately equal to the group velocity of the laser pulse in the plasma $v_w \approx v_g$. This density wave is called the wakefield of the laser pulse. But, even if the laser pulse duration exceeds the plasma wavelength, wakefields can be generated, since the charge separation introduced by the leading edge of the pulse couples back to the laser pulse and vice versa until a so-called self-modulated wakefield is generated (see, e.g., [9]).

Electrons can be trapped in, and accelerated by, the wakefield by being injected onto a rising phase of the plasma wave. They slide down the wave and achieve the maximum kinetic energy once they have reached the valley of the wave. In order to be accelerated to high energies, the electron has to stay on the plasma wave for a long time. For this, a certain injection velocity is necessary; otherwise, it is quickly outrun by the plasma wave. If the laser pulse is strong enough to produce a full modulation of the plasma density, the maximum energy gain E_{max} and the optimum acceleration length of an electron in the wakefield are given by

$$E_{\max} = 4\gamma_{\mathrm{w}}^2 m_0 c^2, \quad l_{\max} \approx \gamma_{\mathrm{w}}^2 \cdot c/\omega_{\mathrm{p}}, \quad \text{with} \quad \gamma_{\mathrm{w}} = \frac{1}{\sqrt{1 - (\omega_{\mathrm{p}}/\omega)^2}} . \quad (2.10)$$

Remark that the Lorentz factor γ_{w} of the plasma wave does depend only on its phase velocity, which is determined by the dispersion relation of the plasma or, in other words, by the ratio of the plasma density to the critical density. In particular, the maximum energy gain of an electron in a wakefield does not depend on the light intensity. The light intensity basically has to provide the plasma density modulation over a long distance. As in the case of the ponderomotive acceleration, electrons enter the wakefield with different velocities and on different phases with respect to the plasma wave. Consequently, the energy gain and the corresponding spectrum of the emitted electrons is broad and even Boltzmann like, so that again a temperature can be attributed to the wakefield accelerated electrons. Typical experimentally achieved energy gains are in the MeV range (see [8, 10] and references therein).

2.3.4 Self-Focussing and Relativistic Channeling

In the discussion so far we have described only the effect of the light field onto the plasma. Because of its dispersion relation, the plasma modifies vice versa the propagation of the laser pulse. We will see that even though complex in detail, the overall effect simplifies and optimizes the electron acceleration process.

The index of refraction of a plasma was given by (2.1):

$$n_{\mathrm{pl}} = \sqrt{1 - \frac{\omega_{\mathrm{pl}}^2}{\omega^2}} = \sqrt{1 - \frac{n_{\mathrm{e}} e^2}{\gamma m_0 \epsilon_0 \omega^2}} . \quad (2.11)$$

In the current context we have to discuss the influence of the electron density n_{e} on the light propagation. We know that the ponderomotive force of the laser beam pushes electrons in radial direction out of the optical axis: a hollow channel is generated along the laser propagation (see Fig. 2.6 [left]). From the numerator in (2.1) we see that the speed of light in the plasma (determined by $v_{\mathrm{p}} = c/n_{\mathrm{pl}}$) increases with increasing electron density. Therefore, the ponderomotively induced plasma channel acts as a positive lens on the laser beam. From the denominator in (2.1) we see that the same effect is caused by the relativistic mass increase of the electrons, which is larger on the optical axis of the beam than in its wings. From their origin these mechanisms are called ponderomotive and relativistic self-focussing, respectively. In competition with the natural diffraction of the beam and further ionization, both effects lead to a filamentation of the laser beam over a distance, which can be much longer than the confocal length (Rayleigh length) of the focussing geometry. This channeling can be beautifully monitored through the nonlinear Thomson scattering of the relativistic electrons in the channel, which is the emission of the figure-of-eight electron motion at harmonics of the laser frequency (see Fig. 2.6 and [11, 12]).

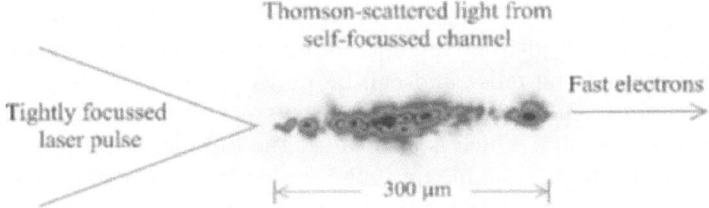

Fig. 2.6. Nonlinear Thomson scattering of relativistic electrons in the relativistic channel. The laser propagates from *left* to *right*, the extension in the displayed example is 300 µm as opposed to the Rayleigh length of 15 µm

2.3.5 Monoenergetic Electrons, the Bubble Regime

In 2004, almost concurrently three groups in the United Kingdom, France, and the United States could demonstrate that high-intensity lasers can produce relativistic and tightly collimated electron beams with a narrow energy spectrum [13, 14, 15] (see Fig. 2.7). This scientific breakthrough opens a wealth of new applications from accelerator physics to nuclear physics.

The mechanism beyond the narrowband acceleration is a subtle interplay between plasma dynamics and intense laser pulse propagation. If the longitudinal extension of the laser pulse is half of the plasma wavelength and if

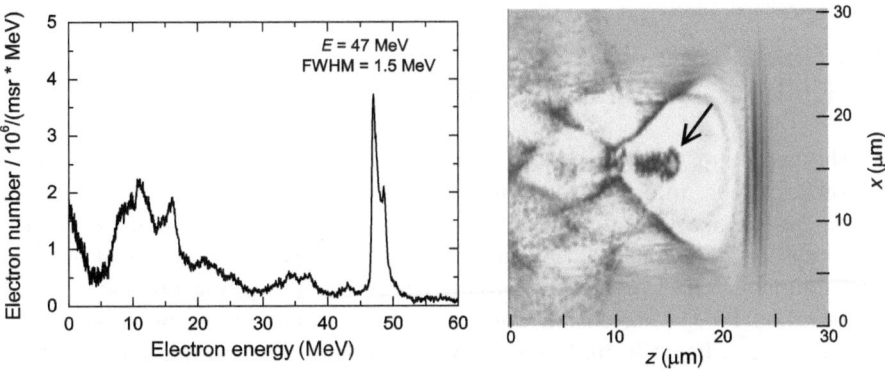

Fig. 2.7. (a) Spectrum of narrowband laser-accelerated relativistic electrons. The narrowband signal at 47 MeV has a spectral width of about 1.5 MeV. It was generated with a 500-mJ, 80-fs laser pulse, tightly focussed with f/2 optic in a subsonic jet of maximum plasma density of $5 \cdot 10^{19}$ cm^{-3}. (b) Three-dimensional particle in cell simulation of electron acceleration in a laser-generated plasma. Propagation direction is from left to right, dark depicts high electron density, and bright low. The large hollow structure is the bubble. All electrons in the center of the bubble (arrow) are accelerated over a length of several hundred micrometers to tens or even hundreds of MeV within a spectral bandwidth of only a few percent. (5 fs, $a_0 = 5$, $w_0 = 5$ µm, (120 mJ), $n_e = 0.01\, n_{cr} = 1.75 \cdot 10^{19}$ cm^{-3}. From ILLUMINATION, M. Geissler

the laser field is strong enough, electrons can be trapped in the wake: electrons that are expelled from the optical axis by the leading edge of the laser pulse flow around the pulse and can be pulled back to the axis by the positive ion charge. This electron motion forms a hollow structure around the highest laser intensity, which is also called the bubble. Some of the electrons can be soaked into the bubble on its back side forming a so-called stem (see Fig. 2.7 [left]). These electrons are accelerated to kinetic energies of tens to even hundreds of MeV. Their energy spread can be less than a few percent and their lateral divergence is only a few mrad. The situation suggests an analogy with a single breaking water wave: As long as the wave amplitude is small, only the phase of the wave propagates with high velocity, and water molecules just oscillate around their rest position. Once the amplitude exceeds the wave breaking threshold, some of the water droplets are trapped in the leading edge of the wave and speed up to the phase velocity of the wave. If the sea ends (in an appropriate shape), the droplets, being the electrons in the plasma wave, can be expelled onto the beach. In case of the electrons the beach is the vacuum, their energies are highly relativistic, and the spectral width is narrow. The situation can be numerically simulated with the help of three-dimensional particle-in-cell codes, in fact it was anticipated before it was experimentally observed [16]. Figure 2.7 shows the typical hollow structure filled with monoenergetic electrons, which evolves in the plasma, if the plasma and laser conditions for bubble acceleration are fulfilled.

2.4 Solid State Targets and Ultrashort Hard X-Ray Pulses

In the preceding chapters, we have described the acceleration of electrons in a homogeneous, underdense plasma by an intense laser pulse. However, electrons can be accelerated also from solid target surfaces. Several mechanisms can cause generation and heating of a plasma and acceleration of electrons and ions near the front of the target, where the prevailing process is mainly determined by the extension and the density gradient of the preplasma at the moment when the main laser pulse incidents. In a simplified view, three regimes can be distinguished [17]. First, if basically no preplasma exists and the laser beam incidents under a finite angle with respect to the target normal, electrons can be accelerated by the normal component of the electric field into the target. Once they have penetrated the solid, they are out of reach of the light field, cannot be drawn back, and deposit their energy inside the matter. This process is called Brunel acceleration or heating [18]. Second, if the preplasma has a considerable extension in front of the solid, and again the light incidents under an angle with respect to target normal, the light can propagate in the preplasma up to the depth, where the plasma density equals the critical density for the laser wavelength and is reflected there. Since here the laser frequency is resonant with the plasma frequency, the light can efficiently couple to the collective plasma oscillation or, in other words, excites a plasma wave pointing along the target normal. This wave is damped in regions of higher density and thereby heats the plasma. Because of the resonant character of the interaction, the mechanism is called resonance absorption and it is the dominating process in most intense laser–solid interaction experiments. Finally, if the preplasma is very long and thin, the acceleration mechanism discussed above as wakefield acceleration and possibly also self-focussing and direct laser acceleration can take place, which again all result in fast electrons in the forward direction (see, e.g., [19]).

When the laser-accelerated electrons are stopped in matter, preferably of high atomic number, ultrashort flashes of Bremsstrahlung are generated. The photon spectrum basically resembles the electron spectrum but it is much more difficult to measure, as will be discussed in the chapter of F. Ewald. Here we will summarize only the main mechanisms of laser generation of Bremsstrahlung.

The basic process of Bremsstrahlung generation is the inelastic scattering of an electron off a nucleus. Since the electron is accelerated when passing the nucleus, it emits radiation. For this elementary process it follows from fundamental arguments that the number of emitted photons per energy interval is constant up to the maximum energy given by the kinetic energy of the electron. However, in a realistic experimental situation the target has a finite thickness. An incident electron successively loses energy in many collisions, where the energy loss per distance strongly depends on its energy: with increasing energy, photo effect, Compton scattering, and finally pair production dominate the interaction, while radiation losses (Bremsstrahlung) always being small. Furthermore, secondary processes have to be included, since the interaction with charged particles dominates: if a fast electron knocks out a bound electron, this secondary electron will again generate Bremsstrahlung and scatter with a third electron and so on until all electrons are stopped.

At nonrelativistic electron energies the angular distribution of the Bremsstrahlung from a thick target is isotropic. However, for relativistic energies the radiation is preferably emitted in forward direction within a cone of decreasing opening angle with energy. On top of this already-complex generation of radiation, the reabsorbtion of photons on their way out of the thick target has to be considered, at least for the low-energy part of the spectrum.

This situation with a manifold of successive processes, each fundamental but complex in total, calls for a Monte Carlo simulation of the situation. In Fig. 2.8a, we show the Bremsstrahlung spectrum of a single electron with 10 MeV energy stopped in a tantalum target ($Z = 72$) of 5-mm thickness. The target is considered as thick since the electron is completely stopped within it. The spectrum was computed by a Monte Carlo–based simulation code (MCNPX [20]), which includes all elementary electron scattering mechanisms, all secondary processes, all relativistic propagation effects, and even photonuclear reactions induced by the Bremsstrahlung (see also [12]). The spectral density

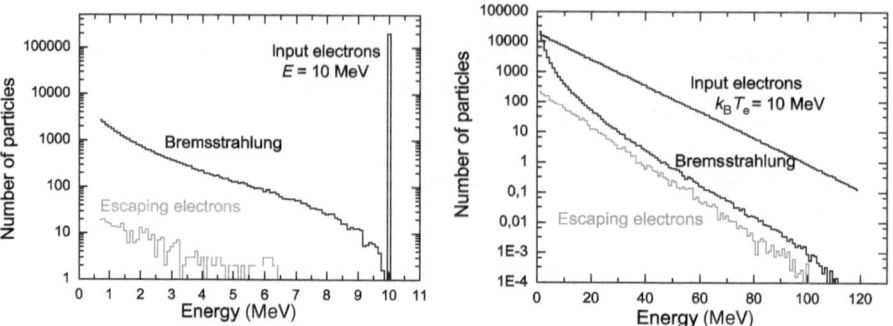

Fig. 2.8. Calculated Bremsstrahlung spectrum from a 5-mm-thick tantalum target obtained from (**a**) a monochromatic electron of 10 MeV and (**b**) an incident exponential electron spectrum with a temperature of $k_B T_e = 10$ MeV. The calculations are performed with MCNPX [20]

drops exponentially with photon energy up to the maximum at the electron energy. The total energy conversion from electron to photon energy is in the order of a percent for the given situation.

In typical laser experiments the incident electrons are far from being monochromatic. Their spectrum is either purely Boltzmann-like at least for energies above 1 MeV or a combination of a Boltzmann-like spectrum and a narrowband component as described above. In order to depict the consequences for the Bremsstrahlung, we plot the photon spectrum obtained from an exponential electron spectrum of $k_B T_e = 10$ MeV incident on the same thick tantalum target as above (see Fig. 2.8b). We see that the resulting photon spectrum is exponential with a slightly lower temperature than of the incident electrons and an overall energy conversion of again a few percent.

These Bremsstrahlung photons are generated by ultrashort electron pulses, which are produced by the ultrashort laser pulse. Therefore, the duration of the photon flash is again ultrashort. It is lengthened only by the time the electrons need to be stopped within the target. The energy of the photons can be considerably higher than the separation energy of neutrons and protons in nuclei and can therefore be applied to induce nuclear processes on an ultrashort timescale of 10^{-12} s or less.

2.5 Proton and Ion Acceleration

Given the situation where the ultrashort and intense laser pulse has accelerated electrons on the front side of a solid target in forward direction, a new phenomenon of directed ion acceleration occurs. The current understanding of the process is as follows (see Fig. 2.9). If the target is thin enough that the ultrashort pulse of fast electrons can exit its back surface, a sheath of negative charge is built up beyond the surface. The range of this field is basically

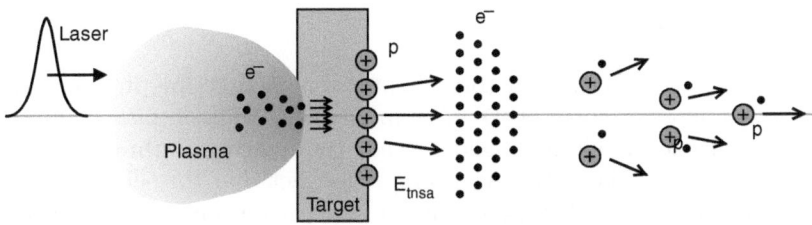

Fig. 2.9. Target Normal Sheath Acceleration: Interaction of a laser pulse with a thin metal foil. Electrons accelerated on the front side penetrate the foil and build up a sheath of high negative charge beyond the back surface. The huge quasi-static electric field between sheath and back surface ionizes and accelerates ions in a highly laminar beam to energies of several MeV per nucleon

limited by the Debye shielding length, and its strength can be as large as TV/m. During this process, ions on the back surface of the target are ionized and then accelerated by the huge quasi-static field. An imperative condition of this process seems to be that the target rear surface may not be melted or destroyed by the time the field is built up. In this case the lightest surface ions are accelerated in a direction normal to the target surface. Therefore, the mechanism is called target normal sheath acceleration of ions [21].

Once a considerable number of ions are accelerated, they compensate the negative charge in the sheath, the field breaks down, and a basically charge compensated cloud of ions and electrons fly straight away. If the target surface is not carefully cleaned, the lightest ions on it are hydrogen, carbon, and oxygen from the rest gas in the vacuum. Predominantly these ions are accelerated. Their spectrum exhibits a slowly decreasing but continuous shape with a sharp cutoff at its high-energy end [22, 23]. The maximum energy for protons can reach tens of MeV in experiments with high-energy lasers as the Vulcan PetaWatt laser and many MeV with tabletop multi-TW lasers [24, 25]. An intriguing property of these ion beams is their extremely low emittance, which has its origin in the very small phase space of the ions in combination with the ultrashort duration of the acceleration [26]. The ions are basically at rest before they are accelerated, and the charge of the ion bunch is compensated by the picked-up electrons after a very short time.

Other than electrons, ions cannot be accelerated to relativistic energies yet and their spectra have no insinuations of monochromaticity as it is the case for electrons. But without doubt, through the success in the laser-based generation of well-collimated, narrowband electron beams, the efforts toward a similar situation with protons are tremendous. First, theoretical and numerical suggestions are circulating, even though probably the next generation of higher intensity lasers has to evolve before this goal can be reached. The current status of ion acceleration and their properties is discussed in detail in the chapters of Victor Malka and Paul McKenna in this book.

2.6 Conclusion

The intention of this introductory chapter was to describe, in simple words, the current understanding of the main mechanisms of laser–matter interaction if the amplitude of the light field exceeds the relativistic threshold for electron acceleration. The author is aware of the fact that the selection of topics is far from being complete and some of the mechanisms also are still under discussion. Some of these discussions will be deepened in the following chapters. But finally all the applications of lasers in nuclear physics first require the up conversion of quantum energy from the 1 eV photon out of the laser to a photon or particle with an energy of many MeV. If you, the reader, have gained a vivid understanding of these fundamental processes, you will be able to appreciate and enjoy the rest of the book.

References

1. C. Danson, P. Brummitt, R. Clarke, J. Collier, B. Fell, A. Frackiewicz, S. Hancock, S. Hawkes, C. Hernandez-Gomez, P. Holligan, M. Hutchinson, A. Kidd, W. Lester, I. Musgrave, D. Neely, D. Neville, P. Norreys, D. Pepler, C. Reason, W. Shaikh, T. Winstone, R. Wyatt, B. Wyborn: Nucl. Fusion **44**(12), S239 (2004). URL: http://stacks.iop.org/0029-5515/44/S239
2. M. Pittman, S. Ferré, J. Rousseau, L. Notebaert, J. Chambaret, G. Chériaux: Appl. Phys. B Lasers Opt. **74**(6), 529 (2002)
3. W. Kruer: *The Physics of Laser Plasma Interactions* (Addison Wesley, 1988)
4. S.C. Wilks, W.L. Kruer, M. Tabak, A.B. Langdon: Phys. Rev. Lett. **69**, 1383 (1992)
5. G. Malka, J.L. Miquel: Phys. Rev. Lett. **77**, 75 (1996)
6. T. Tajima, J.M. Dawson: Phys. Rev. Lett. **43**, 267 (1979)
7. E. Esarey, P. Sprangle, J. Krall, A. Ting: IEEE Trans. Plasma Sci. **24**(2), 252 (1996)
8. F. Amiranoff, S. Baton, D. Bernard, B. Cros, D. Descamps, F. Dorchies, F. Jacquet, V. malka, J. Marques, G. Matthieussent, P. Mine, A. Modena, P. Mora, J. Morillo, Z. Najmudin: Phys. Rev. Lett. **81**, 995 (1998)
9. T. Antonsen, P. Mora: Phys. Rev. Lett. **69**(15), 2204 (1992)
10. F. Amiranoff: Meas. Sci. Technol. **12**, 1795 (2001)
11. C. Gahn, G. Tsakiris, A. Pukhov, J. Meyer-ter Vehn, G. Pretzler, P. Thirolf, D. Habs, K. Witte: Phys. Rev. Lett. **83**, 4772 (1999)
12. B. Liesfeld, K.U. Amthor, F. Ewald, H. Schwoerer, J. Magill, J. Galy, G. Lander, R. Sauerbrey: Appl. Phys. B **79**, 1047 (2004)
13. S. Mangles, C. Murphy, Z. Najmudin, A. Thomas, J. Collier, A. Dangor, E. Divall, P. Foster, J. Gallacher, C. Hooker, D. Jaroszynski, A. Langley, W. Mori, P. Norreys, F. Tsung, R. Viskup, B. Walton, K. Krushelnick: Neature **431**, 535 (2004)
14. C. Geddes, C. Toth, J. van Tilborg, E. Esarey, C. Schroeder, D. Bruhwiler, C. Nieter, J. Cary, W. Leemans: Nature **431**, 538 (2004)
15. J. Faure, Y. Glinec, A. Pukhov, S. Kiselev, S. Gordienko, E. Lefebvre, J.P. Rousseau, F. Burgy, V. Malka: Nature **431**, 541 (2004)

16. A. Pukhov, J. Meyer-ter Vehn: Appl. Phys. B **74**, 355 (2002)
17. P. Gibbon, E. Förster: Plasma Phys. and Contro. Fus. **38**(6), 769 (1996). URL: http://stacks.iop.org/0741-3335/38/769
18. F. Brunel: Phys. Rev. Lett. **59**, 52 (1987)
19. M.I.K. Santala, M. Zepf, I. Watts, F.N. Beg, E. Clark, M. Tatarakis, K. Krushelnick, A.E. Dangor, T. McCanny, I. Spencer, R.P. Singhal, K.W.D. Ledingham, S.C. Wilks, A.C. Machacek, J.S. Wark, R. Allot, R. Clarke, P. A. Norreys: Phys. Rev. Lett. **84**(7), 1459 (2000)
20. J. Hendricks, et al.: MCNPX, version 2.5e, techn. rep. LA-UR 04-0569. Tech. rep., Los Alamos National Laboratory (February 2004)
21. S. Wilks, A. Langdon, T. Cowan, M. Roth, M. Singh, S. Hatchett, M. Key, D. Pennington, A. MacKinnon, R. Snavely: Phys. of Plasmas **8**(2), 542 (2001)
22. M. Hegelich, S. Karsch, G. Pretzler, D. Habs, K. Witte, W. Guenther, M. Allen, A. Blazevic, J. Fuchs, J.C. Gauthier, M. Geissel, P. Audebert, T. Cowan, M. Roth: Phys. Rev. Lett. **89**(8), 085002 (2002). URL: http://link.aps.org/abstract/PRL/v89/e085002
23. M. Kaluza, J. Schreiber, M.I.K. Santala, G.D. Tsakiris, K. Eidmann, J.M. ter Vehn, K.J. Witte: Phys. Rev. Lett. **93**(4), 045003 (2004). URL: http://link.aps.org/abstract/PRL/v93/e045003
24. S. Fritzler, V. Malka, G. Grillon, J. Rousseau, F. Burgy, E. Lefebvre, E. d'Humieres, P. McKenna, K. Ledingham: Appl. Phys. Lett. **83**(15), 3039 (2003)
25. P. McKenna, K.W.D. Ledingham, S. Shimizu, J.M. Yang, L. Robson, T. McCanny, J. Galy, J. Magill, R.J. Clarke, D. Neely, P.A. Norreys, R.P. Singhal, K. Krushelnick, M.S. Wei: Phys. Rev. Lett. **94**(8), 084801 (2005). URL: http://link.aps.org/abstract/PRL/v94/e084801
26. T.E. Cowan, J. Fuchs, H. Ruhl, A. Kemp, P. Audebert, M. Roth, R. Stephens, I. Barton, A. Blazevic, E. Brambrink, J. Cobble, J. Fernandez, J.C. Gauthier, M. Geissel, M. Hegelich, J. Kaae, S. Karsch, G.P.L. Sage, S. Letzring, M. Manclossi, S. Meyroneinc, A. Newkirk, H. Pepin, N. Renard-LeGalloudec: Phys. Rev. Lett. **92**(20), 204801 (2004). URL: http://link.aps.org/abstract/PRL/v92/e204801

3

Laser-Triggered Nuclear Reactions

F. Ewald

Institut für Optik und Quantenelektronik, Friedrich-Schiller-Universität Jena
Max-Wien-Platz 1, 07743 Jena

3.1 Introduction

Nearly 30 ago, laser physicists dreamed of the laser as a particle accelerator [1]. With the acceleration of electrons, protons, and ions up to energies of several tens of MeV by the interaction of an intense laser pulse with matter, this dream has become reality within the last ten years. Today, highly intense laser systems drive microscopic accelerators. Nuclear reactions are induced by the accelerated particles. This article intends to outline the unique properties of laser-based particle and bremsstrahlung sources, and the diversity of new ideas that arise from the combination of lasers and nuclear physics.

Triggering nuclear reactions by a laser is done indirectly by accelerating electrons to relativistic velocities during the interaction of a very intense laser pulse with a laser-generated plasma. These electrons give rise to the generation of energetic bremsstrahlung, when they are stopped in a target of high atomic number. They can as well be used to accelerate protons or heavier ions to several tens of MeV. Those bremsstrahlung photons, protons, and ions with energies in the typical range of the nuclear giant dipole resonances of about a few to several tens of MeV may then induce nuclear reactions, such as fission, the emission of photoneutrons, or proton-induced emission of nucleons. To induce one of these reactions, a certain energy threshold – the activation energy of the reaction – must be exceeded.

Since the first demonstration experiments, nuclear reactions were used for the spectral characterization of laser-accelerated electrons and protons as well as bremsstrahlung [2, 3, 4, 5]. A whole series of classical known nuclear reactions has been shown to be feasible with lasers, such as photo-induced fission [6, 7], proton- and ion-induced reactions [5, 8, 9], or deuterium fusion [10, 11, 12, 13, 14]. Recently the cross section of the (γ,n)-reaction of ^{129}I was measured in laser-based experiments [15, 16, 17].

This last step from the pure observation of nuclear reactions to the measurement of nuclear parameters is of importance regarding the small size of

F. Ewald: *Laser-Triggered Nuclear Reactions*, Lect. Notes Phys. **694**, 25–45 (2006)
www.springerlink.com   © Springer-Verlag Berlin Heidelberg and European Communities 2006

nowaday's high-intensity laser systems compared to large accelerator facilities. It is a first step to a possible joint future of nuclear and laser physics. Nevertheless, all probable future applications of laser-induced nuclear reactions would need to have properties that are not covered by classical nuclear physics. Otherwise, they would stay a diagnostics tool for laser–plasma physicists. The striking properties of a laser as driving device for nuclear reactions are its small tabletop size, the possibility to switch very fast from one accelerated particle to another as well as the ultrashort duration of these particle and bremsstrahlung pulses.

3.2 Laser–Matter Interaction

The basis of all laser-triggered nuclear reactions is the acceleration of particles such as electrons, protons, and ions as well as the generation of high-energy bremsstrahlung photons by the interaction of very intense laser pulses incident on matter. The mechanisms of particle acceleration change sensitively with the target material and chemical phase. The choice of target material in conjunction with the laser parameters is important for the control of plasma conditions and therewith for the control of optimum particle acceleration. Gaseous targets and underdense plasmas are suited best for the acceleration of electrons to energies of several tens of MeV [18, 19, 20, 21]. Thin solid targets, in contrary, are used to accelerate protons and ions [5, 22, 23, 24]. Deuterium fusion reactions have been realized with both heavy water droplets and deuterium-doped plastic [10, 12, 14]. Therefore, but without being exhaustive, the different acceleration mechanisms of electrons, protons, and ions that are important for the production of energetic electrons, protons, and photons are outlined in this section.

3.2.1 Solid Targets and Proton Acceleration

The interaction of short and intense laser pulses with solid targets leads to the formation of a dense plasma that is opaque for the incident laser radiation. This plasma is generated by the rising edge of the incident laser pulse, while the interaction of the highest intensity part of the pulse with this preformed plasma heats and accelerates the plasma electrons. The dominant mechanisms of electron acceleration in such dense plasmas are resonance absorption [25] and ponderomotive acceleration [26]. For laser intensities above $10^{18} \, \text{W/cm}^2 \, \mu\text{m}^2$, where the electron oscillation in the strong electromagnetic light field leads to relativistic electron energies, the mean electron energy, or temperature, of ponderomotively accelerated electrons scales with intensity like [27, 28]

$$k_\text{B} T_\text{hot} = 0.511 \, \text{MeV} \left[\left(1 + \frac{I\lambda^2}{1.37 \cdot 10^{18} \, \text{W/cm}^2 \, \mu\text{m}^2} \right)^{1/2} - 1 \right]. \quad (3.1)$$

The mean electron energies obtained by the interaction of a short laser pulse with solid targets, that is, dense plasmas with a steep density gradient, usually reach several MeV at laser intensities of $10^{19} - 10^{20}$ W/cm^2. Using prepulses may lead to an increase of electron energy.

A fraction η of these electrons that are accelerated by the laser–target interaction enter and traverse the thin solid target. This acceleration of electrons is the first step to the acceleration of protons and ions by a mechanism called *target normal sheath acceleration* (TNSA) [29, 30]: when the electrons leave the few micrometer thin solid target at the rear surface with an electron density n_e and a temperature $k_B T_e$, they leave behind a positively charged target layer. Thus a high electrostatic space-charge field of the order of

$$E \approx k_B T_e/e\lambda_D, \qquad \lambda_D = (\varepsilon_0 k_B T_e/e^2 n_e)^{1/2}, \qquad (3.2)$$

is created, where λ_D is the Debye-length, with ε_0 being the dielectric constant. The electron density $n_e = \eta N_e/(c\tau_L A_F)$ is given by the number of electrons N_e, which are accelerated during a time span given approximately by the duration of the laser pulse τ_L and the focus area A_F. c is the speed of light. $\eta \approx 10 - 20\%$ is the fraction of energy transferred from the laser pulse energy into the electrons that are accelerated and transmitted through the target foil. Therewith from (3.2) it follows that electric fields of about

$$E = \sqrt{(\eta I_L)/(\varepsilon_0 c)} \approx 10^{12} \text{ V/cm} \qquad (3.3)$$

can be generated. This high space-charge field causes field ionization of a few monolayers of rear surface ions, and accelerates them in the direction of the target normal. Under the poor vacuum conditions during laser-matter interaction experiments, these first few monolayers consist of hydrocarbon and water impurities, adsorbed at the target surfaces.

Protons are accelerated first because of their high charge-to-mass ratio. Typical proton spectra (see Fig. 3.1) show an exponential decay with increasing energy followed by a sharp cutoff at energies that depend on the square root of the laser intensity, as can be seen from the above expression for the field strength. This scaling has been proven experimentally for relativistic laser intensities as well as by particle-in-cell simulations [24] and an analytical treatment of the dynamic evolution of the accelerating space-charge field [31]. The cutoff energy as well as the number of acclerated protons depends on both, the intensity and the energy of the laser pulse. Typically, numbers of about 10^9 to 10^{12} protons with a temperature in the order of hundred keV can be accelerated [5, 32, 33]. The maximum energies vary between a few MeV and several tens of MeV. It has been shown in several experiments that the beam quality of laser-accelerated protons can be superior to proton beams from classical particle accelerators with respect to low transversal emittance and a small source size [23]. Only recently it has been demonstrated that even monoenergetic features in the ion spectra can be generated by structuring the target surface [35, 47].

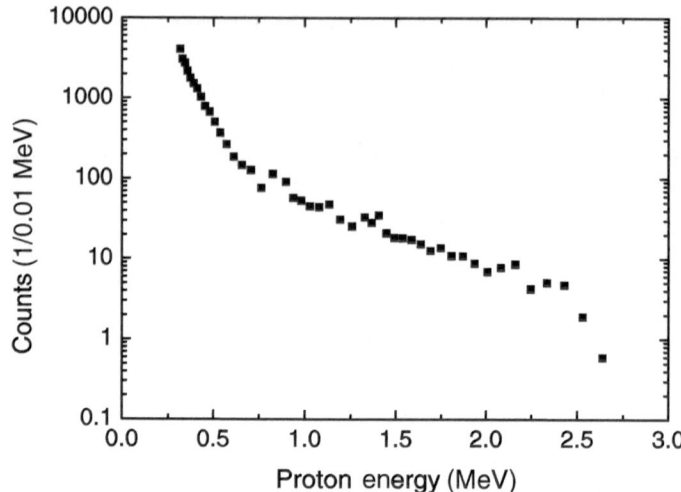

Fig. 3.1. Typical spectrum of protons accelerated by the Jena 15 TW tabletop high-intensity Ti:Sapphire laser (JeTi). The protons were accelerated from a 2-μm-thin tantalum foil with a laser pulse intensity of $I = 6 \times 10^{19}$ W/cm^2. The pulse energy on target was 240 mJ and the pulse duration 80 fs. The number of protons is given in arbitrary units

Protons can also be accelerated from the front side of the target by charge separation-induced fields, but the energies are usually lower [9, 32] and the beam quality is inferior to those of rear surface accelerated protons. Removing the hydrogen-containing contaminants from the target surfaces, for example, by heating of the target [34], leads to the acceleration of ions from the target material itself, such as carbon, fluorine, aluminum, lead, or iron [9, 34, 36, 37]. The observed energies may reach 10 MeV per nucleon while the beam quality is similar to the proton beam.

3.2.2 Gaseous Targets and Electron Acceleration

As we have seen in the previous section, electrons are accelerated because of the interaction of an intense laser pulse with a solid target. The interaction of a laser pulse with a gas may lead to electrons with considerably higher energy and better beam quality. Under certain conditions, intense laser pulses may form self-generated plasma channels in gaseous targets, because of charge separation and relativistic effects [38, 39, 40]. In these channels the intense laser pulse is confined and may be guided over distances about ten times longer than the Rayleigh length of the beam focused in vacuum [41, 42]. In consequence, the high intensity of the focal spot is maintained or even exceeded [40] for several hundreds of microns up to millimeters. Electron acceleration mechanisms such as direct laser acceleration [43], wake field acceleration [38, 44], and the recently investigated regime of forced laser wakefield acceleration [18, 19] or bubble acceleration [45] can therefore act on the plasma electrons over large distances of hundreds of microns or even a few millimeters.

The electric field strength generated by a laser wake, that is, by a laser-excited resonant plasma wave is about 100 GV/cm, allowing for the acceleration of electrons to energies in the order of 100 MeV [18, 19]. When accelerated in the broken wave (or bubble) regime, electrons may be quasi monoenergetic [19, 20, 21, 46].

In the following part of this article, electrons will not be considered as projectiles that induce nuclear reactions. This is due to the fact that even if electrons are able to trigger nuclear reactions, this effect will not be measurable. The cross sections of electron-induced reactions are at least two orders of magnitude smaller than those of photon-induced reactions. Coinstantaneously, electrons that are incident on a solid target will always produce bremsstrahlung. Therefore, photon-induced reactions will always be dominant and electron-induced reactions can be neglected.

3.2.3 Bremsstrahlung

In Sect. 3.3, nuclear reactions induced by laser-generated energetic photons will be reviewed. In particular, the spectrum of bremsstrahlung generated by laser–matter interaction will be used. Therefore, in the following the generation of bremsstrahlung is recalled and the expected photon spectrum is derived.

Photons with energies of up to several tens or hundreds of MeV are generated from the stopping of laser-accelerated electrons inside high-Z materials, such as tantalum, tungsten, or gold. The number of bremsstrahlung photons per energy interval $d(\hbar\omega)$ which are generated by a number of N_e electrons with energy E in a target with number density n is generally given by

$$dn_\gamma(\hbar\omega) = nN_e \frac{d\sigma_\gamma(E)}{d(\hbar\omega)}\, dx\,. \tag{3.4}$$

Therein $d\sigma_\gamma(E)/d(\hbar\omega)$ is – in photon energy – the differential cross section for bremsstrahlung generation. The distance dx, which is travelled by the electrons inside the target material, is given by the stopping power $S = -dE/dx$. In a thick target (with respect to the incident electron energy) where the electrons are stopped completely because of inelastic scattering processes and radiation losses, the path length for the electrons is given by

$$x = \int_{\hbar\omega}^{E_0} \frac{dE}{S}\,, \tag{3.5}$$

where E_0 is the initial electron energy. The lower integration limit follows from the fact that a photon with energy $\hbar\omega$ can be produced only by an electron with at least the same energy value. For high electron energies and thin targets the deflection of the electrons within the target is very small, such that x equals in good approximation the target thickness. In the intermediate range of target thicknesses, the integration has to be stopped at the energy value that the electron possesses when it leaves the target. Insertion of dx and $d\sigma_\gamma(E)/d(\hbar\omega)$ in (3.4) and integration over the electron energy loss yields the number of generated photons per energy interval.

Laser-accelerated plasma electrons are usually not monoenergetic, but show a broad and, in most cases, Boltzmann-like energy spectrum. Therefore (3.4) additionally needs to be integrated over the normalized electron distribution function $f(E_0, T_e)$:

$$dn_\gamma(T_e, \hbar\omega) = nN_e \int_0^\infty f(E_0, T_e) \int_{\hbar\omega}^{E_0} \frac{d\sigma_\gamma(E)}{d(\hbar\omega)} \frac{1}{S} \, dE \, dE_0. \tag{3.6}$$

T_e is thereby the characteristic electron energy or the so-called hot electron temperature. Solving this integral for an exponential spectrum of relativistic electrons yields again an exponential dependence of photon number versus energy in the limit of high photon energies, that is, $\hbar\omega \gg k_B T_e$ [48]. The photon temperature will be lower than the temperature of the incident electrons [48], which has also been proven experimentally, measuring the distributions of laser-accelerated electrons and bremsstrahlung simultaneously [49] (see Fig. 3.2). This behavior can be understood easily, since in an exponentially decaying energy distribution only the few electrons with the highest energies

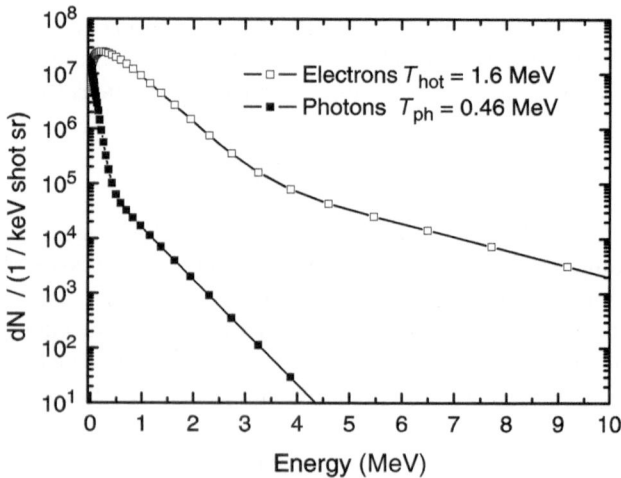

Fig. 3.2. *Upper line*: laser-generated Boltzmann-like electron spectrum from the interaction of a laser pulse of 5×10^{19} W/cm^2 with a solid Ta-target of 2-mm thickness. *Lower line*: simultaneously measured spectrum of Bremsstrahlung. Electrons and photons were measured with spectrometers consisting of thermoluminescence detectors buried in stacks of absorbing metal sheets [49, 50]

can generate energetic photons. And since the number of photons generated by a monoenergetic population of electrons in a thick target diminishes as well with increasing photon energy, the resulting bremsstrahlung distribution will be weighted to lower energies. In addition, in a target with finite thickness, the electrons with the highest energies are able to leave the target. They are lost for the production of energetic bremsstrahlung, which leads to a lowering of the temperature.

In summary, it follows that the high-energy tail of the laser-produced bremsstrahlung spectrum can be described by a Boltzmann distribution

$$n_\gamma(E, T_\gamma)\,\mathrm{d}E = n_0 \cdot e^{-E/k_\mathrm{B}T_\gamma}\mathrm{d}E\,, \tag{3.7}$$

wherein $E = \hbar\omega$, and $n_0 = n_\gamma(E = 0)$ is a normalization constant. In the following sections bremsstrahlung spectra are always presented in this form.

3.3 Review of Laser-Induced Nuclear Reactions

Since the first demonstration experiments laser-induced nuclear reactions have been used to diagnose protons, bremsstrahlung, and (indirectly) electrons, emitted from laser-plasma interactions. They have as well been used to generate neutrons and to demonstrate that it is possible to measure nuclear reaction cross sections. These measurements will be reviewed in the second part of this section, preceded with the recall of the nuclear physics basics needed for these measurements.

3.3.1 Basics of Particle and Photon-Induced Nuclear Reactions and Their Detection

By the impact of energetic photons or particles, such as protons, or ions, the giant dipole resonances (GDR) of nuclei can be excited, resulting in the fission of the nucleus or the emission of nucleons. The particle or photon energies necessary to excite the giant resonances usually lie between a few and several tens of MeVs. These reactions exhibit reaction thresholds that are due to the nucleon binding energy that has to be overcome by the incident particle. Therefore the threshold energies for $(\gamma,2\mathrm{n})$- and $(\gamma,3\mathrm{n})$-reactions are significantly higher (10–30 MeV) than those for the emission of a single neutron (6–10 MeV). Since these reactions are coupled to the excitation of a nuclear resonance, the giant dipole resonance cross sections of spherical nuclei have a nearly Lorentzian shape:

$$\sigma(E) = \sigma_\mathrm{max} \frac{(E\Gamma)^2}{(E^2 - E_\mathrm{max}^2)^2 + (E\Gamma)^2}\,, \tag{3.8}$$

where in E denotes the incident particle (photon) energy, E_max the position of the cross-sectional maximum σ_max, and Γ the full width at half maximum of

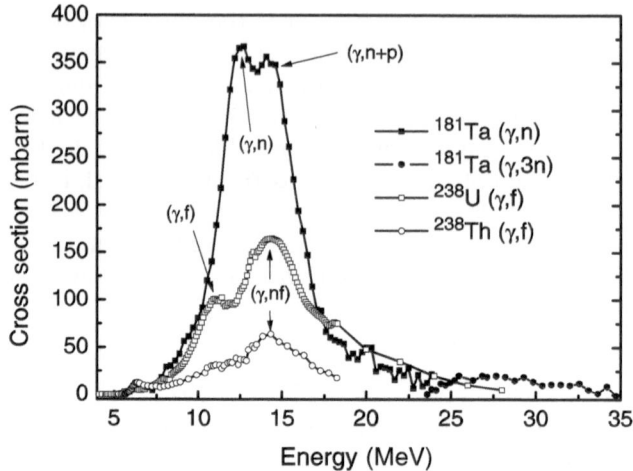

Fig. 3.3. Experimental cross-sectional data for the photo-induced fission of ^{238}U and ^{232}Th [51, 52] as well as the (γ,n) and (γ,3n) cross sections of ^{181}Ta [52]

the resonance. This equation does not account for the reaction barrier, below which the cross section vanishes. In Fig. 3.3, the cross sections of the photon-induced fission of ^{238}U and ^{232}Th as well as for the (γ,n)-reaction of ^{181}Ta are shown as an example.

The number N of nuclear reactions induced, for example, by bremsstrahlung photons normalized to the total number N_0 of irradiated nuclei in a target is

$$N/N_0 = \int_{E_{\text{th}}}^{\infty} \sigma(E) n_\gamma(E,T) \mathrm{d}E \,. \tag{3.9}$$

E_{th} denotes the threshold energy of the reaction, and $n_\gamma(E)$ is the exponential photon distribution given by (3.7).

At the same time the number of induced nuclear reactions in a target irradiated by laser-generated bremsstrahlung can be measured quantitatively, since in many cases the product nuclei of the reactions are radioactive, emitting characteristic γ-rays. These can be detected with high-resolution Ge- or NaI-crystals, allowing for an activity measurement of these lines. The activity of a certain line is given by

$$A = A(t = 0)e^{-t/\tau} \,, \tag{3.10}$$

where the time $t = 0$ denotes the time at which the activation, that is, irradiation of the target is stopped. τ is the mean lifetime of the radioactive isotope, related to the half-life $t_{1/2}$ by $1/\tau = \ln 2/t_{1/2}$. The value that is measured with a γ-detector is the number of decaying nuclei detected during the time $t' = 0$ to t:

$$M(t) = \frac{1}{\varepsilon(E)I_\gamma} \int\limits_{t'=0}^{t} A(t')\mathrm{d}t' = A_0\tau\left[1 - e^{-t/\tau}\right]. \tag{3.11}$$

The parameters $\varepsilon(E)$ and I_γ are the energy-dependent detection efficiency of the detector and the natural abundance of the observed γ-line. The total number N of reactions induced in the target is equal to the number of decays in the time span $t' = 0$ to $t' = \infty$:

$$N = M(t = \infty) = A_0\tau. \tag{3.12}$$

Using this measured value together with (3.9), the incident photon distribution can be derived. But therefore one assumption has to be made: the shape of the photon spectrum has to be known. As has been discussed in Sect. 3.2.3, laser-generated bremsstrahlung spectra can be described by the Boltzmann distribution (3.7). The temperature of this distribution can be derived using two different nuclear reactions, such as the (γ,n)- and the $(\gamma,3\mathrm{n})$-reactions in $^{181}\mathrm{Ta}$. Since the ratio of the number of induced reactions $N_{(\gamma,\mathrm{n})}/N_{(\gamma,3\mathrm{n})}$ can be directly measured through the decay of the product nuclei, the temperature and the number of photons n_0 can be obtained by solving

$$\frac{N_{(\gamma,\mathrm{n})}}{N_{(\gamma,3\mathrm{n})}} = \frac{\int \sigma_{\gamma,\mathrm{n}}(E)\, n_\gamma(E)\, \mathrm{d}E}{\int \sigma_{\gamma,3\mathrm{n}}(E)\, n_\gamma(E)\, \mathrm{d}E}. \tag{3.13}$$

3.3.2 Photo-Induced Reactions: Fission (γ,f), Emission of Neutrons $(\gamma,x\mathrm{n})$, and Emission of Protons (γ,p)

It sounds very spectacular to be able to fission nuclei with the help of intense laser light. But this is indeed what happened for the first time roughly 6 years ago: $^{238}\mathrm{U}$ was fissioned by laser-generated bremsstrahlung [6, 7]. These proofs of principal experiments were the key events, which attracted attention to laser-induced nuclear reactions, although already 2 years before (γ,n)-reactions in copper, gold, and aluminum were used to measure bremsstrahlung emission from laser–matter interactions [53, 54]. Apart from the fission of $^9\mathrm{Be}$, which has a reaction threshold of 1.7 MeV, the photo fission cross sections of $^{238}\mathrm{U}$ and $^{232}\mathrm{Th}$ exhibit the lowest threshold energies of all known photo-induced reactions and were therefore among the first candidates for the demonstration of laser-induced reactions. While fission of $^9\mathrm{Be}$ could be achieved already with some $10^{18}\,\mathrm{W/cm^2}$ [55], for the fission of uranium a laser intensity of at least $10^{19}\,\mathrm{W/cm^2}$ is necessary.

In the mean time it has been shown that in principle all types of photo-induced reactions can be realized by laser-generated bremsstrahlung. Those are mainly the (γ,f)-, $(\gamma,x\mathrm{n})$-, and (γ,p)-reactions (with $x = 1,2,3,\ldots$). Whether a particular reaction can be realized and detected depends on the threshold energy with respect to the bremsstrahlung energies, on the photon flux, and on the cross-sectional value.

Photo-Induced Nuclear Reactions as a Diagnostic Tool for Laser–Plasma Interactions

In all high-intensity laser–matter interaction experiments that produce energetic electrons, material of the target assembly and target chamber may be activated through nuclear reactions [53, 59]. In turn, nuclear activation is used as a diagnostic tool for laser-accelerated energetic electrons and bremsstrahlung [60]. Besides thermoluminescence detector–based spectrometers [49, 50], the activation measurement is the only technique that allows for the measurement of the whole photon spectrum of single MeV-photon bursts with durations below a microsecond.

In the most simple case the activation of one single isotope is observed. The ratio of induced nuclear reactions N to the total number of irradiated nuclei N_0 is given by (3.9). Since the distribution has three free parameters – the shape, the temperature T_γ, and the amplitude – two or more reactions with different threshold energies should be induced and detected at the same time, such as the pair $^{181}Ta(\gamma,n)^{180}Ta$ and $^{181}Ta(\gamma,3n)^{178}Ta$ with threshold energies at 7.6 and 22.1 MeV, respectively. In this case the assumption of an exponential energy distribution is sufficient to calculate the temperature and amplitude of the spectrum using (3.13) [61, 62, 63]. The more reactions are available, and the wider the range of threshold energies, the more accurately the photon spectrum can be reconstructed. From the use of several different reaction thresholds it has been shown that under certain plasma conditions, that is, acceleration mechanisms, the photon spectra follow indeed rather two-temperature distributions than single exponential decays when observed over a large energy regime [7, 61]. In Table 3.1, some nuclear reactions for bremsstrahlung diagnostics are listed together with their reaction parameters such as threshold energy, resonance energy, and cross section maximum. The threshold energies span a wide range of about 8 up to 50 MeV, which allow for the measurement of bremsstrahlung spectra over this same large energy regime. The lower energy limit for this nuclear activation technique is set by the lowest reaction thresholds, which are provided by the fission reactions of ^{238}U and ^{232}Th with roughly 6 MeV or ^9Be with 1.7 MeV.

Angular distributions of bremsstrahlung may be determined similarly: a number of activation samples, such as pieces of copper, are arranged around the laser focus [2, 3, 60, 61, 64]. After irradiation, the activities of these samples are measured. In this way a number of photons emitted within the solid angles covered by the size of each of the samples are deduced. The angular resolved bremsstrahlung and electron distributions can give a direct insight into the plasma properties and electron acceleration mechanisms [64].

Cross-Sectional Measurements

When the energy distribution and the angular spread of the laser-generated bremsstrahlung are characterized with the above-described technique, this information can be used to deduce the cross section from other photo-induced

Table 3.1. Parameters of some photo-induced reactions that have been induced by laser-generated bremsstrahlung. The cross-sectional data are taken from [56, 57] and the decay parameters from [58]. Only the strongest γ-lines of the decaying reaction products are indicated. The ^{180}Ta-, ^{232}Th-, and ^{238}U-cross sections exhibit double peaks, for both of which the cross-sectional values are given

Target Nucleus	Reaction	Product Nucleus	$t_{1/2}$	γ-Line (keV)	E_{th} (MeV)	E_{max} (MeV)	σ_{max} (mbarn)	FWHM (MeV)
^{181}Ta	$(\gamma,1n)$	^{180}Ta	8.152 h	93.3	7.6	12.8	221	2.1
						14.9	330	5.2
^{181}Ta	$(\gamma,3n)$	^{178m}Ta	2.36 h	426.38	22.1	27.7	21	5.6
^{197}Au	$(\gamma,1n)$	^{196}Au	6.18 d	355.68	8.1	13.5	529	4.5
^{197}Au	$(\gamma,3n)$	^{194}Au	38.0 h	328.45	23.1	27.1	14	6.0
^{63}Cu	$(\gamma,1n)$	^{62}Cu	9.7 min	1172	10.8	16	68	8
^{63}Cu	$(\gamma,2n)$	^{61}Cu	3.3 h	282	19.7	25	13.6	6.5
^{65}Cu	$(\gamma,1n)$	^{64}Cu	12.7 h	1345	9.9	16.7	77.5	5
^{64}Zn	$(\gamma,1n)$	^{63}Zn	38 min	669	9.9	16.7	71.8	13
^{238}U	(γ,f)	Fission Prod.	–	–	6.0	14.34	175	8.5
						11.39	113.1	
^{232}Th	(γ,f)	Fission Prod.	–	–	5.8	14.34	63.93	7.0
						6.39	12.44	

reactions. Such a measurement, using laser-generated bremsstrahlung, has been shown recently [15, 16, 17, 63] by the measurement of the cross section maximum of the (γ,n)-reaction in the isotope ^{129}I. Although the values obtained for σ_{max} vary strongly and the errors are still large, these measurements demonstrate that laser-generated energetic bremsstrahlung as well as laser-accelerated particles can be used to measure nuclear reaction parameters quantitatively.

Two techniques of measuring the maximum cross-sectional value were realized: First, σ_{max} is obtained from (3.9) if the parameters of the bremsstrahlung distribution are known from previous or simultaneous nuclear activation measurements and if some assumptions on the cross section are made [15, 63]. As shown in (3.8), the shape of the cross section is Lorentzian-like. The reaction threshold is given by the energy balance of the (γ,n)-reaction. The FWHM of the resonance is the only value which is not known, but can be assumed to be about 5 MeV. From such a measurement the cross section σ_{max} was deduced to be 250 mbarn with an error of about 100 mbarn, which is mostly due to uncertainties in the determination of the photon temperature, which in turn relies on the available cross-sectional data.

A second experiment used the direct comparison of induced (γ,n)-reactions in the isotopes ^{129}I and ^{127}I, which both are contained in the sample [17]. The latter reaction cross section is known such that the ratio of induced reactions corrected for the percent per weight in the sample yields directly the ratio of the integral cross section. With the same assumptions on the shape and FWHM of the unknown ^{129}I cross section as above the cross section maximum has been measured to be 97 ± 40 mbarn. This technique is advantageous, since it is independent from a separate measurement of the bremsstrahlung distribution, which is reflected in a more accurate value.

Another, more approximative, estimate of a reaction cross section was made of the unknown (γ,p)-reaction in ^{58}Ni [61]. Since the reaction cross section for the (γ,n)-reaction in ^{58}Ni is known, the comparison of the number of induced (γ,n)- and (γ,p)-reactions in the same sample gives an approximate value of the energy-integrated cross section of the (γ,p)-reaction.

3.3.3 Reactions Induced by Proton or Ion Impact

Proton-Induced Reactions

Other than in the case of bremsstrahlung, protons and other ions can alternatively be measured by pure magnetic or by Thompson parabola spectrometers. But these cover only a very small solid angle of the emitted particle beam, and can therefore not be used to measure overall numbers of emitted ions, nor can angular resolved measurements be performed easily. Alternatively, an activation technique can be used, similar to the derivation of the bremsstrahlung spectrum by photo-induced nuclear reactions. Some of the most suitable proton-induced nuclear reactions are listed in Table 3.2. As the counting of the emitted characteristic γ-spectrum of the decaying product nuclei is done off-line after the shot, this nuclear activation technique is insensitive to the high electromagnetic pulse that tends to saturate all electronic diagnostics during the time of laser–plasma interaction. Nevertheless, the activation technique is limited to high proton fluxes per shot like proton emission generated by high-energy laser systems ($E \sim 50$ J). For low-energy laser pulses ($\sim$1 J) many shots have to be accumulated to overcome the detection threshold, set by the number of induced reactions and the efficiency of the γ-counting system.

Reaction cross sections for proton-induced reactions can be up to one order of magnitude higher than photon-induced reactions, and as can be seen in Table 3.2 the reaction thresholds are significantly lower, in particular for elements with low atomic number.

With a stack of thin copper foils – in the range of tens to hundreds of microns – laser-produced proton spectra have been measured by the activation of the foils through the ^{63}Cu(p,n)^{63}Zn reaction [5, 65]. The spectral range which is covered by this technique depends on the thickness of the stack, since the spectrum is calculated from the number of reactions induced in a certain depth, that is, in a certain foil. Having measured this number by γ-spectrometry, the number of protons stopped in this foil as well as their energy can be derived from the known cross section and proton-stopping data. The

Table 3.2. Some proton-induced reactions. Cross-sectional and decay data are taken from [57] and from [58], respectively. Only the strongest γ-lines of the decaying reaction products are indicated

Target Nucleus	Reaction	Product Nucleus	$t_{1/2}$	γ-Line (keV)	E_{th} (MeV)	E_{max} (MeV)	σ_{max} (mbarn)
^{65}Cu	(p,n)	^{65}Zn	244.3 d	1115.5	2.13	10.9	760
^{65}Cu	(p,p+n)	^{64}Cu	12.7 h	1345.8	9.91	25	490
^{63}Cu	(p,n)	^{63}Zn	38.47 min	669.6	4.15	13	500
^{63}Cu	(p,2n)	^{62}Zn	9.186 h	596.56	13.26	23	135
^{63}Cu	(p,p+2n)	^{61}Cu	3.33 h	656.0	19.74	40	323
^{11}B	(p,n)	^{11}C	20.39 min	511	3	8	300
^{13}C	(p,n)	^{13}N	9.96 min	511	3	6	150

upper energy detection limit is given only by the number of induced reactions in the stack and the sensitivity of the gamma-detector, whereas the limit for the detection of low-energy protons is given by the reaction threshold of 4 MeV.

Another technique that involves only one single copper foil takes advantage of the different proton-induced reactions in the natural occuring isotopes ^{65}Cu and ^{63}Cu, which are listed in Table 3.2 [5]. Some of these reaction cross sections are plotted in Fig. 3.4. Again, from the known cross-sectional data and the measured number of induced reactions the proton spectrum is derived. Since the cross sections of these reactions peak at energies from 10 to 40 MeV, and the lowest threshold energy is 2 MeV, proton spectra in the range of 2–40 MeV can be measured.

Other proton-induced reactions have been demonstrated as well. In particular the (p,n)-reactions in ^{11}B, ^{13}C, and ^{18}O [65] are of potential interest for future applications in the production of isotopes for positron emission tomography (PET) as is discussed in Sect. 3.4.

Ion-Induced Nuclear Reactions

As shown in Sect. 3.2, ions heavier than protons can be accelerated by the same mechanism that leads to protons acceleration. ^{12}C, ^{16}O – both of which originate from hydrocarbon and water contaminations on the target surfaces – as well as ions from the target material itself, such as ^{27}Al, ^{56}Fe [9, 34, 37], or deuterons [12], were accelerated. It has been shown that the ion energies can reach 10 MeV/nucleon, which results in 650 MeV energy for Fe-ions [9]. These ions can react with a secondary target: they may fuse and form highly excited compound nuclei, which evaporate neutrons, protons, and alpha-particles [9, 37]. Depending on their initial excitation, that is, incident ion energy the reaction channel may be different, corresponding to a different cross section

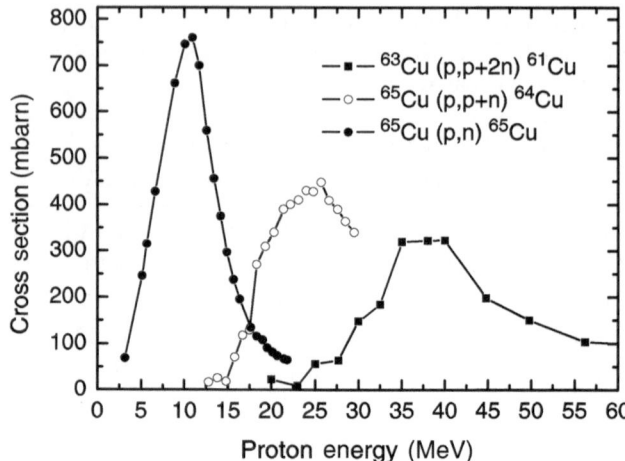

Fig. 3.4. Some cross sections of proton-induced reactions. Experimental data are taken from [57]

and product nucleus. From the fusion of accelerated ^{56}Fe-ions with ^{12}C nuclei 17 fusion–evaporation reaction channels with cross section thresholds in the range of $1.2\,\text{MeV} \leqslant E_{\text{th}} \leqslant 80\,\text{MeV}$ were detected through the decay of their product nuclei [9]. An energy spectrum of the accelerated Fe-ions was derived from this measurement, analogous to the above-discussed proton spectrum measurements.

Deuterium–Deuterium Fusion

By (γ,n)- and (p,n)-reactions neutrons are produced. The spectra of these neutrons are broad, and the number of neutrons is of the same order as the number of nuclear reactions. These nuclear reactions therefore may provide a short-pulsed point-like neutron source.

Another technique to produce neutrons is the fusion of two deuterium nuclei within a laser plasma:

$$D + D \longrightarrow {}^3\text{He} \ (0.82\,\text{MeV}) + \text{n} \ (2.45\,\text{MeV}). \tag{3.14}$$

These fusion neutrons are thus monoenergetic, with a width that is determined by the center-of-mass velocity of the reacting deuterons. The cross section for this reaction increases for deuteron energies above 5 keV and starts saturating at about 50 keV. Ion temperatures of about 100 keV and thus fusion can be achieved in the interaction of an intense laser pulse with deuterated clusters [13, 66], heavy water droplets [14], or deuterated solid plastic targets [10].

Since the fusion reaction takes place only for about the time the plasma is heated by the fs-laser pulse, the duration of neutron emission is very short. The source size is given by the expanding heated plasma, which will be of the order of tens to hundreds of microns in diameter. The generated numbers of neutrons vary between 10^4 [13, 14] and 10^8 [11] per laser shot and depend on the target material as well as on laser energy and intensity.

3.4 Future Applications

Many ideas for potential applications of laser-driven nuclear particle sources and laser-driven nuclear reactions have been developed [67] since the first basic demonstration experiments were carried out. Some of these ideas rely on the compactness and flexibility of the laser–plasma tool and therefore might promise quantitative advantages over conventional methods. Others rely on one or more of the unique properties of the laser particle and radiation sources, which might lead to open access to presently unresolved scientific problems.

Laser-driven fusion of heavier nuclei has already been demonstrated [9, 37] and could be developed in future to measure the fusion cross sections rather than using them for derivation of the ion spectra. As the peak intensity of laser systems continues to increase, laser-driven fusion reactions might also become an efficient method to produce nuclear isotopes and isomers far away from the line of stability.

Protons may induce spallation (p,xn) and fission (p,f) reactions when targeted to nuclei such as lead or uranium. These reactions release a high number of neutrons. These neutron pulses have a very short duration, the source size may be small, and this laser driven neutron production can be stopped instantaneously at any time. Because of these properties, such a neutron source might prove to have a variety of applications, such as time-resolved neutron radiography.

The properties of a laser-produced plasma, that is, ionization state, high density, and high ion temperature, can be very similar to the conditions of matter present in stars. This plasma environment influences nuclear reaction rates by many factors such as screening effects and strong electromagnetic fields. It is, for example, well known that lifetimes of nuclear levels that involve electrons from the atomic electron cloud, such as electron capture, internal conversion, and $\beta^{+/-}$-decay, may be altered by the ionization state or the influence of strong electric fields [68]. These are effects that will occur during nucleosynthesis in stars. In the laboratory these extreme plasma conditions are available only in a laser-produced plasma. Using this plasma as target and laser-accelerated particles as projectiles, the dependence of nuclear reaction rates on plasma conditions may be investigated. The results will improve the knowledge of the nuclear reaction rates used as inputs in the astrophysics codes that aim at reproducing the evolution of the stars. Improving the reliability of these codes, that is, of the input parameters is one of the major issues in astrophysics today.

Among the main applications that are discussed and developed currently is the laser-driven production of radio isotopes for medical purposes. Radioactive isotopes, such as ^{11}C, ^{13}N, ^{18}F, ^{128}I, and ^{99}Tc, are commonly used in medical imaging such as PET or scintigraphy. Pharmaceutical carriers are tagged with the radioactive isotopes and then injected in the human body. There, they accumulate selectively in certain parts of the body, such as tumors, the thyroid, or bones, where the carrier molecule is used. Thus these parts of the body are marked with the radioactive isotopes. The emitted characteristic γ-photons are detected. PET takes advantage of the fast annihilation of the positrons emitted by short-lived β^+-active isotopes. The 511 keV annihilation photons, emitted under 180° to each other, are used for 3D imaging of the marked region [69]. For medical purposes, short-lived isotopes are favorable, since the activity is high, such that lower quantities of the isotope can be applied and the exposure of the patient to ionizing radiation is reduced. Such short-lived isotopes are usually produced by energetic proton beams from cyclotrons or van-de-Graaff accelerators via (p,n)- and (p,α)-reactions. Some isotopes could be produced as well by (γ,n)-reactions, but proton induced reactions are favored, because reactant and produced isotope have different atomic numbers, which allow for fast chemical separation of the radioactive tracer product. In addition, as mentioned above, the cross sections for proton-induced reactions are higher than those induced by γ-photons.

Because of the size, cost, and shielding required for cyclotrons, the production of PET-isotopes is limited to only a few facilities. The isotopes have to be transported to the hospital over long distances. Thus, for a 3-h transit the amount of initial activity of a short-lived isotope with, for example, 30 min half-life, has at least to be $2^6 = 64$ times higher than the finally used activity. The tabletop laser systems, which could be used for isotope production, can be as small as a few square meters, and shielding is reduced to the nearest vicinity of the laser focus and the target. It is aimed to develop turnkey laser systems with smaller size, but high repetition rate as well as increased intensity and energy for optimized proton acceleration. With such a system a hospital can use laser-driven nuclear activation on-site, which would save costs and facilitate logistics.

It has been shown experimentally with a high-power, high-energy single-shot laser system ($\sim$100 J, $I \approx 10^{20}$ W/cm^2) that activities of 200 kBq from the isotope ^{11}C can be induced in a single laser shot via the ^{11}B(p,n)^{11}C reaction [32] whereas a tabletop laser (1 J, 5×10^{19} W/cm^2) has shown to be able to generate ^{11}C activities of 134 kBq after 30 min of laser irradiation with a pulse repetition rate of 10 Hz [33]. When laser shots are accumulated, saturation of the induced activity will set in after some time because of simultaneous decay of the produced isotope. For a typical 10 Hz high-intensity tabletop laser the integrated activities saturate at 209 kBq for ^{11}C and at 170 kBq for ^{18}F [33]. Hence, with current tabletop lasers typical medical doses of about 800 MBq [32] are out of reach [33]. Thus, the proton flux and the repetition rate of the laser have to be increased. Augmenting the latter from 10 Hz to 1 kHz would allow to achieve ^{11}C-activities in the order of GBq [33], which would be in the range of realistic applications.

If the first steps of lasers in the field of nuclear science should be followed by real applications, laser-driven nuclear physics has to prove competitiveness with classical accelerators and with nuclear physics instruments and techniques. Or it has to explore complementary techniques, such as the examination of nuclear dynamics in an extreme plasma environment. The some unique properties of laser-accelerated particle and bremsstrahlung radiation, such as the short pulse duration, high particle flux, and in the case of protons the excellent emittance of the source, open new fields to nuclear physics.

The possibility of generating particles, such as electrons, protons, heavier ions, and neutrons, as well as bremsstrahlung, just by changing the target material provides a versatile but nevertheless small accelerator system.

Higher photon and particle fluxes, as well as monoenergetic particle beams, are needed for most future applications. Optimization of the target and plasma parameters and exploiting new acceleration schemes for these particles will be essential. The performance of the existing laser systems has to be augmented, or rather to be adapted to particle acceleration. Currently, powerful, but relatively small laser systems are designed and partly under construction [70]. The aim of these recent efforts in the development of high-intensity lasers is to combine high focused intensities with reasonable high pulse energies (at least a few Joules) and a high repetition rate of which the latter is indispensable for applications such as the medical isotope production.

The above-presented applications still present a challenge. But provided, the laser development and the exact preparation and control of target and plasma properties can be ensured for the coming years, high-intensity laser-induced nuclear physics will have a high probability of entering medical, technical, and basic nuclear physics applications.

References

1. T. Tajima, J. Dawson: Phys. Rev. Lett. **43**, 267 (1979)
2. W. Leemans, D. Rodger, P. Catravas, C. Geddes, G. Fubiani, E. Esarey, B. Chadwick, R. Donahue, A. Smith: Phys. Plasmas **8**, 2510 (2001)
3. G. Malka, M. Aléonard, J. Chemin, G. Claverie, M. Harston, J. Scheurer, V. Tikhonchuk, S. Fritzler, V. Malka, P. Balcou, G. Grillon, S. Moustaizis, L. Notebaert, E. Lefebvre, N. Cochet: Phys. Rev. E **66**, 066402 (2002)
4. H. Schwoerer, F. Ewald, R. Sauerbrey, J. Galy, J. Magill, V. Rondinella, R. Schenkel, T. Butz: Europhys. Lett. **61**, 47 (2003)
5. J.M. Yang, P. McKenna, K.W.D. Ledingham, T. McCanny, S. Shimizu, L. Robson, R.J. Clarke, D. Neely, P.A. Norreys, M.S. Wei, K. Krushelnick, P. Nilson, S.P.D. Mangles, R.P. Singhal: Appl. Phys. Lett. **84**, 675 (2004)
6. K. Ledingham, I. Spencer, T. McCanny, R. Singhal, M. Santala, E. Clark, I. Watts, F. Beg, M. Zepf, K. Krushelnick, M. Tatarakis, A. Dangor, P. Norreys, R. Allot, D. Neely, R. Clark, A. Machacek, J. Wark, A. Cresswell, D. Sanderson, J. Magill: Phys. Rev. Lett. **84**, 899 (2000)

7. T. Cowan, A. Hunt, T. Phillips, S. Wilks, M. Perry, C. Brown, W. Fountain, S. Hatchett, J. Johnson, M. Key, T. Parnall, D. Pennington, R. Snavely, Y. Takahashi: Phys. Rev. Lett. **84**, 903 (2000)
8. R. Snavely, M. Key, S. Hatchett, T. Cowan, M. Roth, T. Phillips, M. Stoyer, E. Henry, T. Sangster, M. Singh, S. Wilks, A. MacKinnon, A. Offenberger, D. Pennington, K. Yasuike, A. Langdon, B. Lasinski, J. Johnson, M. Perry, E. Campbell: Phys. Rev. Lett. **85**(14), 2945 (2000)
9. P. McKenna, K. Ledingham, J. Yang, L. Robson, T. McCanny, S. Shimizu, R. Clarke, D. Neely, K. Spohr, R. Chapman, R. Singhal, K. Krushelnick, M. Wei, P. Norreys: Phys. Rev. E **70**, 036405 (2004)
10. G. Pretzler, A. Saemann, A. Pukhov, D. Rudolph, T. Schätz, U. Schramm, P. Thirolf, D. Habs, K. Eidmann, G.D. Tsakiris, J. Meyer-ter Vehn, K.J. Witte: Phys. Rev. E **58**, 1165 (1998)
11. P. Norreys, A. Fews, F. Beg, A. Dangor, P. Lee, M. Nelson, H. Schmidt, M. Tatarakis, M. Cable: Plasma Phys. Control. Fusion **40**, 175 (1998)
12. K. Nemoto, A. Maksimchuk, S. Banerjee, K. Flippo, G. Mourou, D. Umstadter, V. Bychenkov: Appl. Phys. Lett. **78**(5), 595 (2001)
13. G. Grillon, P. Balcou, J.P. Chambaret, D. Hulin, J. Martino, S. Moustaizis, L. Notebaert, M. Pittman, T. Pussieux, A. Rousse, J.P. Rousseau, S. Sebban, O. Sublemontier, M. Schmidt: Phys. Rev. Lett. **89**, 065005 (2002)
14. S. Karsch, S. Düsterer, H. Schwoerer, F. Ewald, D. Habs, M. Hegelich, G. Pretzler, A. Pukhov, K. Witte, R. Sauerbrey: Phys. Rev. Lett. **91**(1), 015001 (2003)
15. J. Magill, H. Schwoerer, F. Ewald, J. Galy, R. Schenkel, R. Sauerbrey: Appl. Phys. B **77**, 387 (2003)
16. F. Ewald, H. Schwoerer, S. Düsterer, R. Sauerbrey, J. Magill, J. Galy, R. Schenkel, S. Karsch, D. Habs, K. Witte: Plasma Phys. Control. Fusion **45**, A83 (2003)
17. K. Ledingham, P. McKenna, J. Yang, J. Galy, J. Magill, R. Schenkel, J. Rebizant, T. McCanny, S. Shimizu, L. Robson, R. Singhal, M. Wei, S. Mangles, P. Nilson, K. Krushelnick, R. Clarke, P. Norreys: J. Phys. D: Appl. Phys. **36**, L63 (2003)
18. V. Malka, S. Fritzler, E. Lefebvre, M.M. Aleonard, F. Burgy, J.P. Chambaret, J.F. Chemin, K. Krushelnick, G. Malka, S.P.D. Mangles, Z. Najmudin, M. Pittman, J.P. Rousseau, J.N. Scheurer, B. Walton, A.E. Dangor: Science **298**, 1598 (2002)
19. J. Faure, Y. Glinec, A. Pukhov, S. Kiselev, S. Gordienko, E. Lefebvre, J.P. Rousseau, F. Burgy, V. Malka: Nature **431**, 541 (2004)
20. C. Geddes, C. Toth, J. van Tilborg, E. Esarey, C. Schroeder, D. Bruhwiller, C. Nieter, J. Cary, W. Leemans: Nature **431**, 538 (2004)
21. S. Mangles, C. Murphy, Z. Najmudin, A. Thomas, J. Collier, A. Dangor, E. Divall, P. Foster, J. Gallacher, C. Hooker, D. Jaroszynski, A. Langley, W. Mori, P. Norreys, F. Tsung, R. Viskup, B. Walton, K. Krushelnick: Nature **431**, 535 (2004)
22. M. Kaluza, J. Schreiber, M.I.K. Santala, G.D. Tsakiris, K. Eidmann, J. Meyerter Vehn, K.J. Witte: Phys. Rev. Lett. **93**, 045003 (2004)
23. T. Cowan, J. Fuchs, H. Ruhl, A. Kemp, P. Audebert, M. Roth, R. Stephens, I. Barton, A. Blazevic, E. Brambrink, J. Cobble, J. Fernández, J.C. Gauthier, M. Geissel, M. Hegelich, J. Kaae, S. Karsch, G.P. Le Sage, S. Letzring, M. Manclossi, S. Meyroneinc, A. Newkirk, H. Pépin, N. Renard-LeGalloudec: Phys. Rev. Lett. **92**, 204801 (2004)

24. A. Maksimchuk, K. Flippo, H. Krause, G. Mourou, K. Nemoto, D. Shultz, D. Umstadter, R. Vane, V.Y. Bychenkov, G.I. Dudnikova, V.F. Kovalev, K. Mima, V.N. Novikov, Y. Sentoku, S.V. Tolokonnikov: Plasma Phys. Rep. **30**, 473 (2004)
25. W.L. Kruer: *The Physics of Laser Plasma Interaction* (Addison-Wesley, Redwood-City, 1988)
26. G. Malka, J.L. Miquel: Phys. Rev. Lett. **77**, 75 (1996)
27. S.C. Wilks, W.L. Kruer, M. Tabak, A.B. Langdon: Phys. Rev. Lett. **69**, 1383 (1992)
28. P. Gibbon, E. Förster: Plasma Phys. Control. Fusion **38**, 769 (1996)
29. S. Hatchett, C. Brown, T. Cowan, E. Henry, J. Johnson, M. Key, J. Koch, A. Langdon, B. Lasinski, R. Lee, A. Mackinnon, D. Pennington, M. Perry, T. Philipps, M. Roth, T. Sangster, M. Singh, R. Snavely, M. Stoyer, S. Wilks, K. Yasuike: Phys. Plasmas **7**, 2076 (2000)
30. S.C. Wilks, A.B. Langdon, T.E. Cowan, M. Roth, M. Singh, S. Hatchett, M.H. Key, D. Pennington, A. MacKinnon, R.A. Snavely: Phys. Plasmas **8**, 542 (2001)
31. P. Mora: Phys. Rev. Lett. **90**, 185002 (2003)
32. I. Spencer, K. Ledingham, R. Singhal, T. McCanny, P. McKenna, E. Clark, K. Krushelnick, M. Zepf, F. Beg, M. Tatarakis, A. Dangor, P. Norreys, R. Clarke, R. Allott, L. Ross: Nucl. Instr. Meth. Phys. Res. B **183**, 449 (2001)
33. S. Fritzler, V. Malka, G. Grillon, J.P. Rousseau, F. Burgy, E. Lefebvre, E. d'Humières, P. McKenna, K. Ledingham: Appl. Phys. Lett. **83**, 3039 (2003)
34. M. Hegelich, S. Karsch, G. Pretzler, D. Habs, K. Witte, W. Guenther, M. Allen, A. Blazevic, J. Fuchs, J.C. Gauthier, M. Geissel, P. Audebert, T. Cowan, M. Roth: Phys. Rev. Lett. **89**(8), 085002 (2002)
35. B.M. Hegelich, B.J. Albright, J. Cobble, K. Flippo, S. Letzring, M. Paffett, H. Ruhl, J. Schreiber, R.K. Schulze, J.C. Fernandez, Nature **439**, 441–444 (2006)
36. E. Clark, K. Krushelnick, M. Zepf, F. Beg, M. Tatarakis, A. Machacek, M. Santala, I. Watts, P. Norreys, A. Dangor: Phys. Rev. Lett. **85**, 1654 (2000)
37. P. McKenna, K. Ledingham, T. McCanny, R. Singhal, I. Spencer, M. Santala, F. Beg, K. Krushelnick, M. Tatarakis, M. Wei, E. Clark, R. Clarke, K. Lancaster, P. Norreys, K. Spohr, R. Chapman, M. Zepf: Phys. Rev. Lett. **91**, 075006 (2003)
38. E. Esarey, P. Sprangle, J. Krall, A. Ting: IEEE Transact. Plasma Sci. **24**, 252 (1996)
39. A. Pukhov, J. Meyer-ter Vehn: Phys. Rev. Lett. **76**, 3975 (1996)
40. P. Gibbon, F. Jakober, A. Monot, T. Auguste: IEEE Transac. Plasma Sci. **24**, 343 (1996)
41. G. Sarkisov, V. Bychenkov, V. Novikov, V. Tikhonchuk, A. Maksimchuk, S.Y. Chen, R. Wagner, G. Mourou, D. Umstadter: Phys. Rev. E **59**, 7042 (1999)
42. R. Fedosejevs, X. Wang, G. Tsakiris: Phys. Rev. E **56**, 4615 (1997)
43. A. Pukhov, Z. Sheng, J. Meyer-ter Vehn: Phys. Plasmas **6**, 2847 (1999)
44. F. Amiranoff, S. Baton, D. Bernard, B. Cros, D. Descamps, F. Dorchies, Jacquet, V. Malka, J.R. Marques, G. Matthieussent, P. Mine, A. Modena, P. Mora, J. Morillio, Z. Najmudin: Phys. Rev. Lett. **81**, 995 (1998)
45. A. Pukhov, J. Meyer-ter Vehn: Appl. Phys. B **74**, 355 (2002)
46. B. Hidding, K.-U. Amthor, B. Liesfeld, H. Schwoerer, S. Karsch, M. Geissler, L. Veisz, K. Schmid, J.G. Gallacher, S.P. Jamison, D. Jaroszynski, G. Pretzler, R. Sauerbrey, Physical Review Letters **96**, 105004 (2006)
47. H. Schwoerer, S. Pfotenhauer, O. Jäckel, K.-U. Amthor, B. Liesfeld, W. Ziegler, R. Sauerbrey, K.W.D. Ledingham, T. Esirkepov, Nature **439**, 445–448 (2006)

48. G. McCall: J. Phys. D: Appl. Phys. **15**, 823 (1982)
49. R. Behrens, H. Schwoerer, S. Düsterer, P. Ambrosi, G. Pretzler, S. Karsch, R. Sauerbrey: Rev. Sci. Instr. **74**(2), 961 (2003)
50. R. Behrens, P. Ambrosi: Radiat. Protect. Dosim. **104**, 73 (2002)
51. J. Caldwell, E. Dowdy, B. Berman, R. Alvarez, P. Meyer: Phys. Rev. C **21**, 1215 (1980)
52. Centre for Photonuclear Experiments Data, Lomonosov Moscow State University. URL: http://depni.sinp.msu.ru/cdfe/muh/calc_thr.shtml. Online Database
53. M. Key, M. Cable, T. Cowan, K. Estabrook, B. Hammel, S. Hatchett, E. Henry, D. Hinkel, J. Kilkenny, J. Koch, W. Kruer, A. Langdon, B. Lasinski, R. Lee, B. MacGowan, A. MacKinnon, J. Moody, J. Moran, A. Offenberger, D. Pennington, M. Perry, T. Phillips, T. Sangster, M. Singh, M. Stoyer, M. Tabak, M. Tietbohl, K. Tsukamoto, K. Wharton, S. Wilks: Phys. Plasmas **5**, 1966 (1998)
54. T.W. Phillips, M.D. Cable, T.E. Cowan, S.P. Hatchett, E.A. Henry, M.H. Key, M.D. Perry, T.C. Sangster, M.A. Stoyer: Rev. Sci. Instr. **70**, 1213 (1999)
55. H. Schwoerer, P. Gibbon, S. Düsterer, R. Behrens, C. Ziener, C. Reich, R. Sauerbrey: Phys. Rev. Lett. **86**, 2317 (2001)
56. IAEA, IAEA-TECDOC (2000)
57. Experimental Nuclear Reaction Data (EXFOR / CSISRS). URL: http://www.nndc.bnl.gov/exfor/exfor00.htm. Online Database
58. The Lund/LBNL Nuclear Data Search. URL: http://nucleardata.nuclear.lu.se. Online Database
59. M. Perry, J. Sefcik, T. Cowan, S. Hatchett, A. Hunt, M. Moran, D. Pennington, R. Snavely, S. Wilks: Rev. Sci. Instr. **70**, 265 (1999)
60. P. Norreys, M. Santala, E. Clark, M. Zepf, I. Watts, F. Beg, K. Krushelnick, M. Tatarakis, X. Fang, P. Graham, T. McCanny, R. Singhal, K. Ledingham, A. Creswell, D. Sanderson, J. Magill, A. Machacek, J. Wark, R. Allot, B. Kennedy, D. Neely: Phys. Plasmas **6**, 2150 (1999)
61. M. Stoyer, T. Sangster, E. Henry, M. Cable, T. Cowan, S. Hatchett, M. Key, M. Moran, D. Pennington, M. Perry, T. Phillips, M. Singh, R. Snavely, M. Tabak, S. Wilks: Rev. Sci. Instr. **72**, 767 (2001)
62. I. Spencer, K.W.D. Ledingham, R.P. Singhal, T. McCanny, P. McKenna, E.L. Clark, K. Krushelnick, M. Zeph, F.N. Beg, M. Tatarakis, A.E. Dangor, R.D. Edwards, M.A. Sinclair, P.A. Norreys, R.J. Clarke, R.M. Allot: Rev. Sci. Instr. **73**, 3801 (2002)
63. B. Liesfeld, K.U. Amthor, F. Ewald, H. Schwoerer, J. Magill, J. Galy, G. Lander, R. Sauerbrey: Appl. Phys. B **79**, 419 (2004). DOI DOI:10.1007/s00340-004-1637-9
64. M. Santala, M. Zepf, I. Watts, F. Beg, E. Clark, M. Tatarakis, K. Krushelnick, A. Dangor, T. McCanny, I. Spencer, R.P. Singhal, K. Ledingham, S. Wilks, A. Machacek, J. Wark, R. Allot, R. Clarke, P. Norreys: Phys. Rev. Lett. **84**(7), 1459 (2000)
65. M. Santala, M. Zepf, F. Beg, E. Clark, A. Dangor, K. Krushelnick, M. Tatarakis, I. Watts, K. Ledingham, T. McCanny, I. Spencer, A. Machacek, R. Allott, R. Clarke, P. Norreys: Appl. Phys. Lett. **78**, 19 (2001)
66. J. Zweiback, R. Smith, T. Cowan, G. Hays, K. Wharton, V. Yanovsky, T. Ditmire: Phys. Rev. Lett **84**(12), 2634 (2000)

67. K. Ledingham, P. McKenna, R. Singhal: Science **300**, 1107 (2003)
68. F. Bosch, T. Faestermann, J. Friese, F. Heine, P. Kienle, E. Wefers, K. Zeitelhack, K. Beckert, B. Franzke, O. Klepper, C. Kozhuharov, G. Menzel, R. Moshammer, F. Nolden, H. Reich, B. Schlitt, M. Steck, T. Stöhlker, T. Winkler, K. Takahashi: Phys. Rev. Lett. **77**, 5190 (1996)
69. K. Kubota: Ann. Nucl. Med. **15**, 471 (2001)
70. J. Hein, S. Podleska, M. Siebold, M. Hellwing, R. Bödefeld, G. Quednau, R. Sauerbrey, D. Ehrt, W. Wintzer: Appl. Phys. B **79**, 419 (2004)

4

POLARIS: An All Diode-Pumped Ultrahigh Peak Power Laser for High Repetition Rates

J. Hein, M. C. Kaluza, R. Bödefeld, M. Siebold, S. Podleska, and
R. Sauerbrey

Institut für Optik und Quantenelektronik, Friedrich-Schiller-Universität,
Max-Wien-Platz 1, 07743 Jena, Germany
jhein@ioq.uni-jena.de

4.1 Introduction

After the invention of the technique of chirped pulse amplification (CPA) [1] we have witnessed a tremendous progress in laser development over the last years. Nowadays, pulses having peak powers in the terawatt (TW) regime can be produced using table-top laser systems that easily fit into university-scale laboratories and operate at repetition rates of 10 Hz and more. These pulses can be focused to reach peak intensities in excess of $10^{19}\,\mathrm{W/cm^2}$ on target which enables us to study the physics of relativistic laser-plasma interaction. However, in order to generate pulses with peak powers of 1 petawatt (PW) and more and intensities beyond $10^{21}\,\mathrm{W/cm^2}$ one still has to use large-scale facilities delivering pulses containing energies of a few 10's J or a couple of 100's J depending on the laser material used. Due to cooling issues these PW-laser systems can be operated at a shot rate of 1 to 3 shots per hour only. This severely limits the variety and complexity of experiments that can be carried out with such laser systems. An increasing number of envisaged applications – such as laser-driven proton beam generation [2, 3], which might be a future alternative in cancer treatment or the laser-induced isotope production for medical applications [4] or a short-wavelength X-ray source based on laser-accelerated electrons [5] – require a high averaged flux of the accelerated particles and consequently a high repetition rate of the laser system driving the initial acceleration process. In addition, parametrical studies varying as many experimental parameters as possible are required to study the physics underlying the acceleration processes [6], also calling for higher shot rates during the experiment. At the Institute of Optics and Quantum Electronics of the University of Jena, Germany, the PW-laser system POLARIS (Petawatt Optical Laser Amplifier for Radiation Intensive Experiments), which is entirely diode-pumped is presently under development. After commissioning it

J. Hein et al.: *POLARIS: An All Diode-Pumped Ultrahigh Peak Power Laser for High Repetition Rates*, Lect. Notes Phys. **694**, 47–66 (2006)
www.springerlink.com   © Springer-Verlag Berlin Heidelberg and European Communities 2006

will be able to deliver PW-laser pulses at a repetition rate of 0.03 or 0.1 Hz. POLARIS will enable us to carry out experiments in a parameter regime, which could not be reached by any other laser system so far. In this article, the underlying technology and the basic design of POLARIS are reviewed. Furthermore, based on the prospect to use recently developed technologies for solid-state laser systems, a scenario for PW-laser systems operating at repetition rates of several Hz is discussed.

At the moment, all operational PW-class systems rely on high-energy flash lamp-pumped Nd:glass lasers [7, 8, 9]. One approach is to use the Nd-doped glass itself as the active medium to amplify the chirped fs-pulses. The minimal pulse duration is then restricted to at least 350 fs due to limitations of the amplification bandwidth. Thus, at least 350 J of pulse energy is required to reach the PW-level. Alternatively, frequency-doubled Nd:glass lasers delivering ns-pulses containing more than 50 J can be used to pump large-diameter Ti:Sapphire crystals that can amplify pulses with a significantly wider bandwidth. Using this amplification scheme, the pulses can be as short as 33 fs containing up to 28 J [9]. However, as all these PW-systems rely on large-scale Nd:glass lasers, the shot rate – determined by the cooling time of the flash lamps and the amplifier discs in the Nd:glass laser – is in the range of 1–3 shots per hour. In this article, we will discuss three aspects to improve the high-energy laser technology to increase the repetition rate and the overall performance of the laser: the use of laser diodes as the pump source for the amplifiers, the laser material according to the pump wave length, and a new compressor design.

As all solid-state lasers are optically pumped a high-efficiency and high-brightness source of pump photons is required to optimize the laser performance. The most efficient light source we presently know is the diode laser. The electrical to optical efficiency can be as high as 74%. The width of the emission spectrum of a high-power laser diode is typically less than 3 nm, which fits very well to the absorption band of solid-state laser materials. However, the beam profile emitted from a laser diode is not cylindrically symmetric, but it has two different divergencies determined by the optical properties of the laser diode itself. This behavior has to be taken into account for an optimised design of the pump geometry. Furthermore, when the pump source consists of a large number of diode bars, pump-beam homogenization will become an important issue.

Once it has been decided to choose laser diodes as the pump source, a suitable laser material has to be found. It is of course possible to stick to all the sophisticated Nd:glass amplifier developments and simply replace the flash lamps by laser diodes. Due to the more efficient pumping by diodes, significantly less heat is deposited in the laser material, which would allow for a repetition rate at least one order of magnitude higher. Admittedly, laser diodes come at much higher costs than flash lamps. Therefore designs that entail a lower loss of pump photons are preferred.

Choosing a well-suited laser material strongly depends on both pump source properties and laser performance goals. For a maximum peak power yield the laser emission bandwidth should be as large as possible to support the amplification of laser pulses as short as possible. In addition, the fluorescence lifetime is a more important issue in diode-pumped lasers than for the flash lamp-pumped case. The reason is that the flash lamp pulse can be made short enough to match the lifetime of the excited neodymium. Moreover, adding another flashlamp to the amplifier does not drive the budget significantly. In the case of diode pumping the peak power of the pump source is limited by the number of diodes. Increasing the pump pulse duration while maintaining a constant power will increase the number of available pump photons for a constant budget as long as these photons can be stored in the amplifier for a sufficiently long time. Ytterbium as the active laser ions perform better under these considerations because typically the upper-state lifetime is longer as well as emission and absorption bandwidths are broader compared to Neodymium. Although the saturation fluence of the laser medium will increase rendering the energy extraction from the amplifier more difficult, shorter pulses can be amplified, and the requirement for the emission bandwidth of the diodes is less strict.

For POLARIS, an Ytterbium-doped fluoride phosphate glass was chosen as the active material. As a quasi-three-level laser material the quantum defect and accordingly the deposited heat is small. Diode lasers emitting at 940 nm are a well-suited pump source and can be operated in pulsed mode matching the 1.5-ms fluorescence lifetime. A good overlap of bandwidth and wavelength shift to the absorption band can be achieved. Efforts in imaging the output of the diode stack to the laser material have been made to ensure a homogeneous pump beam profile and a good spatial overlap. In the following paragraphs we will show that the laser fluence in the amplifiers has to be pushed to its limits for efficiency reasons. In order to handle a flux as high as possible a careful design and high-quality optical components are required for this kind of laser system.

Like in all other CPA systems, a final recompression of the pulses after their amplification is necessary. If the peak power is increased the diameter of the laser beam and therefore the size of the optical gratings and of the vacuum compressor chamber have to grow accordingly due to damage threshold issues. In addition, the required grating size depends on the duration of the stretched pulse, which has to be maximized to avoid damage in the amplifier chain. Hence, the maximum output power depends on the available grating size. Tiling the compressor gratings may overcome this limitation [1]. The demands on the accuracy of the grating alignment in all dimensions as well as possible techniques to ensure this accuracy will be discussed in the last section of this article.

4.2 Ytterbium-Doped Fluoride Phosphate Glass as the Laser Active Medium

One of the most crucial issues when designing a diode-pumped high peak power laser system is the choice of the gain material. Because of the limited peak power at which laser diodes at their present stage can be operated a long fluorescence lifetime for energy storage is desired. For longer fluorescence lifetimes either the emission cross section, or the gain bandwidth, or both decrease. For efficient amplification the energy density of the extracted laser pulse has to be as close to the saturation fluence of the gain medium as possible. In order to achieve a high peak power at moderate pulse energies a large amplification bandwidth for short-pulse generation with CPA system is required. Figure 4.1a illustrates the possibilities for generating high peak powers using some established laser materials. The inverse of the product of saturation fluence and shortest pulse duration indicates the capability of a laser material used to be for high amplification at maximum bandwidth. Assuming gaussian-shaped gain spectra, the correlation between emission cross section, bandwidth, and fluorescence lifetime of a laser material is given by [11]

$$\sigma_{\mathrm{em}} = \frac{c_0^2}{4\pi n^2 \nu^2} \frac{1}{\tau_{\mathrm{f}}} \frac{\sqrt{\ln 2}}{\Delta\nu\sqrt{\pi}} \tag{4.1}$$

with c_0 being the speed of light in vacuum, n the refractive index, h Planck's constant, ν the center frequency, τ_{f} the fluorescence lifetime, and $\Delta\nu$ the bandwidth (FWHM). Applying the time–bandwidth product for a gaussian line-shape, a criterion for the generation of high peak powers at the corresponding fluorescence lifetime depending on the laser wavelength λ and the refractive index n is obtained [12]

$$\frac{\tau_{\mathrm{f}}}{t_{\mathrm{p}} F_{\mathrm{sat}}} \leq 4.26 \cdot 10^9 \frac{(\lambda/\mu\mathrm{m})^3}{n^2} \, \mathrm{cm}^2 \, \mathrm{J}^{-1}. \tag{4.2}$$

Here the saturation fluence is given by $F_{\mathrm{sat}} = h\nu/\sigma_{\mathrm{em}}$, t_{p} is the pulse duration.

In Fig. 4.1b, suitable materials for amplification to high energy levels are mapped. Furthermore, there is an optimum region between high gain and low gain where either amplified spontaneous emission or damage of the laser medium are likely to occur.

Fluoride phosphate glasses are suitable hosts for Yb ions to form a laser material [13]. Their preparation, structure, and properties depending on the composition are very well characterized [14]. Their potential was first shown with continuous wave lasers [15, 16]. But, at the same time the broad emission bandwidth allows for the generation of ultra-short pulses [17].

An advantage of fluoride phosphate glasses compared to phosphates is the thermal behavior [18]. Fluoride phosphate exhibits a negative thermal refractive index change that partially compensates self-focusing effects inside

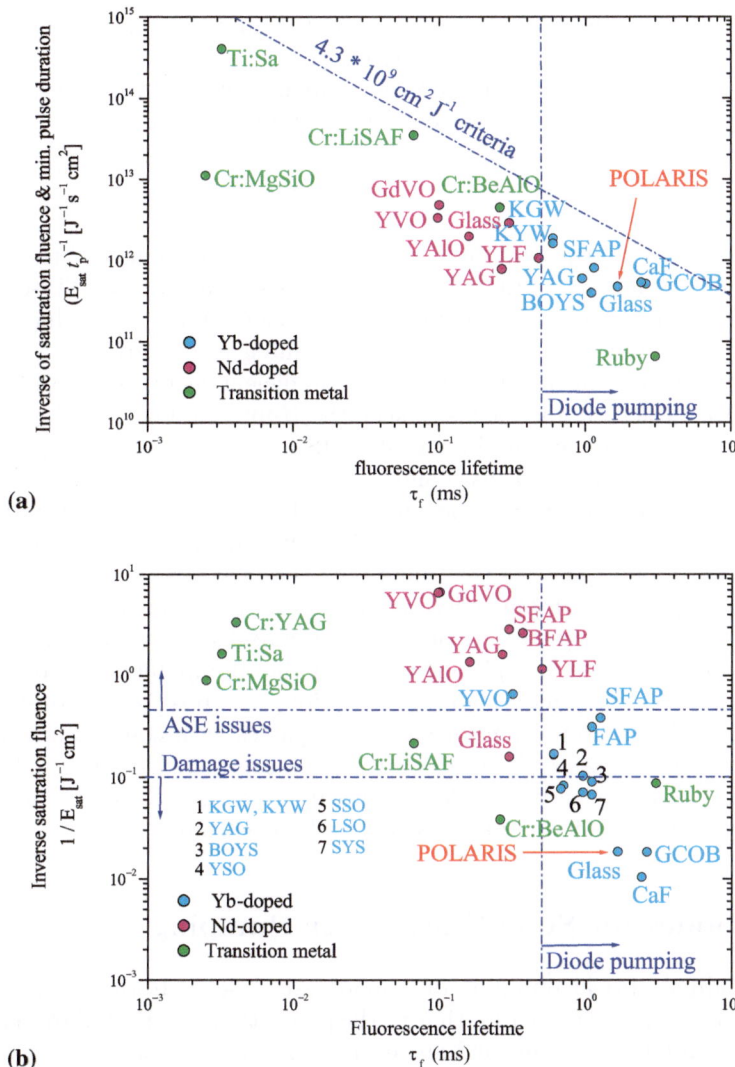

Fig. 4.1. (a) Capability of laser materials for energy storage and generation of high peak power. The inverse saturation fluence multiplied by minimum possible pulse width plotted against fluorescence lifetime τ_f. Differently doped materials are marked with different colors. (b) Capability of laser materials for energy storage and generation of high-energy pulses. The inverse saturation fluence is plotted against the fluorescence lifetime. For a higher gain, amplified spontanous emission (ASE) becomes an issue whereas for lower gain damage limits efficient energy extraction. Diode pumping will be useful if the fluorescence lifetime exceeds a certain limit. The limits are marked with *slash-dot* lines

amplifiers. This was shown by time-resolved pump probe measurements at the pump and laser wavelength, respectively [19].

An important design parameter for a high-power CPA laser system is the B-integral, which describes the collected nonlinear phase

$$B = \int k\, n_2\, I\, \mathrm{d}l, \tag{4.3}$$

where $k = 2\pi/\lambda$ is the wave number, l the propagation length, I the intensity, and n_2 the nonlinear refractive index. In order to avoid intensity-dependent aberrations, the value of B should be kept below unity. The reduction of intensity inside the laser amplifier chain is limited as already discussed by the stretching factor and the fluence needed for efficient energy extraction. A small nonlinear refractive index helps to keep the B-integral low. The fluoride phosphate glasses which are used for POLARIS have $n_2 = 2 \times 10^{-16}\,\mathrm{cm^2/W}$ [20], which is 75% of the n_2 of fused silica.

The glass discs used in the amplifiers have to be polished and surface treated by ion beam etching just before coating them using ion beam sputtering. It was found that this treatment enhances the surface damage threshold considerably. Using an antireflective coating on the laser material low polarization independent losses are achieved for zero-degree incidence. Avoiding a design using Brewster's angle allows for a higher pump fluence by polarization coupling of the laser diode light. In addition, it is possible to rotate the polarization of the laser beam for successive passes with convenient separation by thin film polarizers. Coatings on the glass are designed for a high damage threshold and a reflectivity of less than 0.01% for the laser center wavelength $\pm 20\,\mathrm{nm}$ and 0.2% for the pump wavelength at 940 nm.

4.3 Diodes for Solid State Laser Pumping

Longitudinal optical pumping of a solid state laser material requires a light source with a high brightness, that is, the power that is delivered from a certain source area into a certain solid angle. In addition, if the dopant concentration is fixed, the required brightness depends on the total amplifier output power. This power influences the beam size, whereas the material length corresponds to the absorption length. The result of this consideration is that larger amplifiers require less brightness in the pump source, which holds in a situation where the energy density is fixed.

Nevertheless, the brightness from a laser diode bar can be increased only by increasing the current and consequently the total output power. However, this is limited because higher currents have a strong impact on the diodes' lifetime. If more than one diode bar has to be used as a pump source, the bars should be placed in space as dense as possible in order to avoid a decrease of brightness in the pump area. If they have to be less densely distributed in space for any reason, optical beam-steering techniques may transform the

Power supply
   Laser diode bars (25)
     Directing prisms (3)

Redirecting
prims (25)

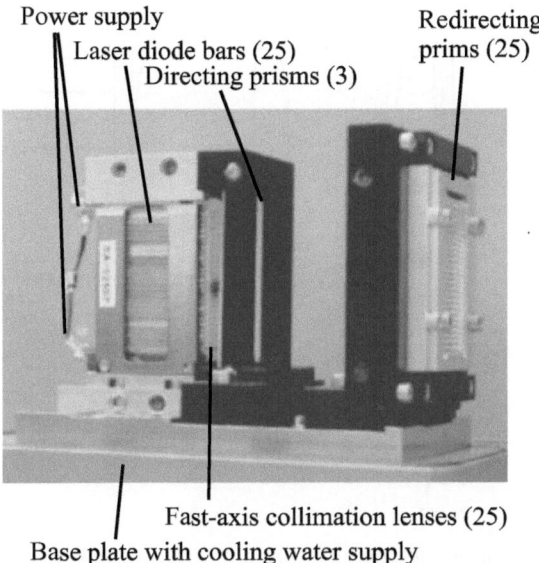

Fast-axis collimation lenses (25)
Base plate with cooling water supply

Fig. 4.2. Laser diode stack used for pumping the POLARIS laser assembled of 25 diode bars with fast axis collimation lenses mounted directly to the heat sinks and a patented beam shaping optics

brightness back to the original value. Reasons for a less-dense distribution are cooling requirements as well as diode and collimation lens mounting needs. One solution is a diode stack as shown in Fig. 4.2. This configuration is used in the POLARIS system.

High average power laser diode production for continuous wave application is well established. A 1-cm wide InGaAsP/InGaP diode bar with 500 single emitters typically delivers 50 W. For pulsed mode operation with duty cycle below 0.01 the peak power can be doubled by increasing the emitter density on the semiconductor bar to 90%.

The diode stacks including the housing used for POLARIS were originally developed for fiber coupling [21]. A patented [22] beam shaping device illustrated in Fig. 4.3 splits up the beam from all 25 bars into three parts by a three prism array which addidionally overlaps them in space. The individual beams are then redirected by an array of 75 small prisms. Therefore the gaps inside the diode stack beam caused by the presence of heat sinks are filled with diode light, and the brightness is increased by a factor of 3. All these components are integrated in a housing that is hermetically sealed. The output from the diode stack assembly is then imaged to the laser material. The exact optical arrangements for the individual amplifier stages are described below.

If the laser diodes are driven by a pulsed instead of continuous current mode the maximum current allowed without significantly stressing their

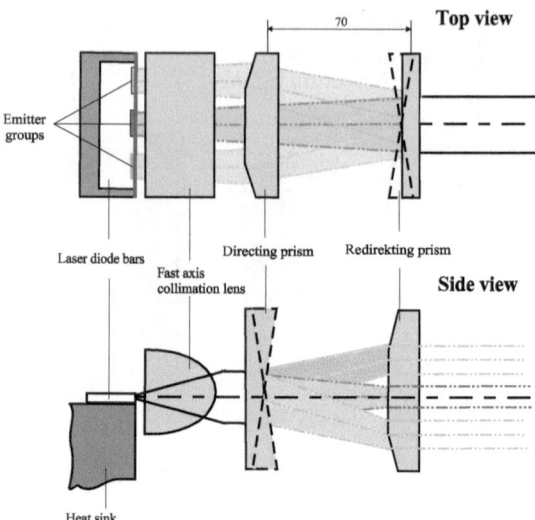

Fig. 4.3. Illustration of the Jenoptik patent [22] for pump beam shaping. There is only one bar shown

lifetime can be increased by a factor of up to 2. The diodes used for POLARIS are driven by rectangular 2.6-ms long pulses with an amplitude of 150 A. It is well known that the emitted wavelength of a laser diode changes with its temperature. For 940-nm GaAs lasers the wavelength shift is about 0.3 nm/K. That results in a wavelength shift across the current pulse. This is shown in Fig. 4.4, where the time-dependent spectrum of the output is measured. The resulting integrated spectrum is also shown as well as the spectrally integrated output power. These data have to be used for the laser design.

4.4 The POLARIS Laser

The POLARIS laser consists of a series of amplifiers fed by a low-energy ultrashort pulse front end. The pulse energy increases at each amplification stage to reach about 150 J in front of the petawatt compressor. The outline of the POLARIS System is shown in Fig. 4.5.

The front end consists of the pulse source, the commercial laser system Mira 900 pumped by a 10 W-Verdi, and the pulse stretcher. The Ti:Sapphire oscillator is tuned to a center wavelength of 1042 nm. At an average output power of 300 mW a 76-MHz train of pulses each having a bandwidth of 20 nm is generated.

These pulses are then stretched to 2.2 ns by a grating stretcher in which the pulses pass the grating 8-times with a hard clip bandwidth of 32 nm. The stretcher incorporates a 14-inch grating with 1480 lines/mm. A clipping

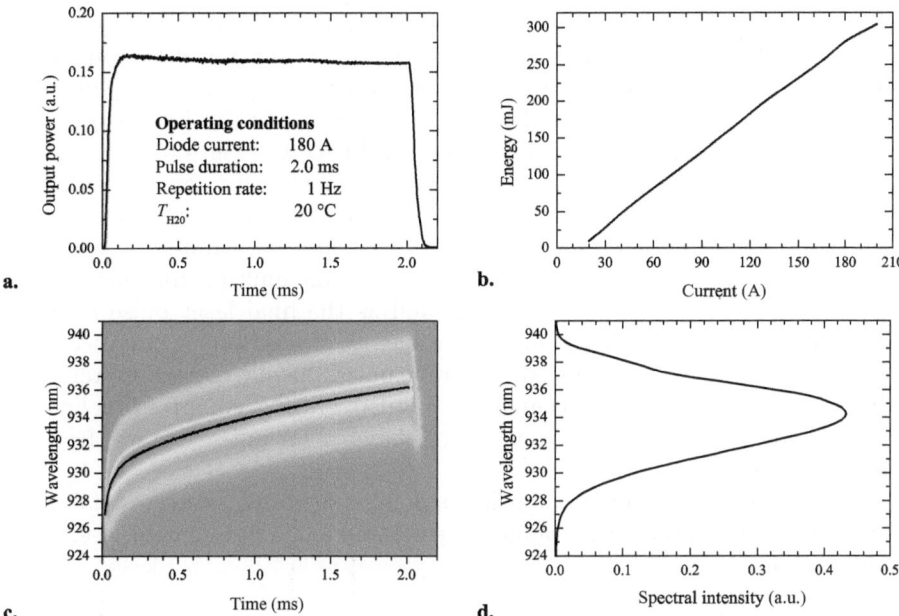

Fig. 4.4. Wavelength shift of a pulsedly driven laserdiode. (**a**) Output power versus time (**b**) Laser diode output energy versus driving current for a 2-ms current pulse. (**c**) Time resolved output spectrum of the diode for a curent pulse of 2 ms with 180 A amplitude (**d**) Time integrated spectrum for the 180 A current pulse

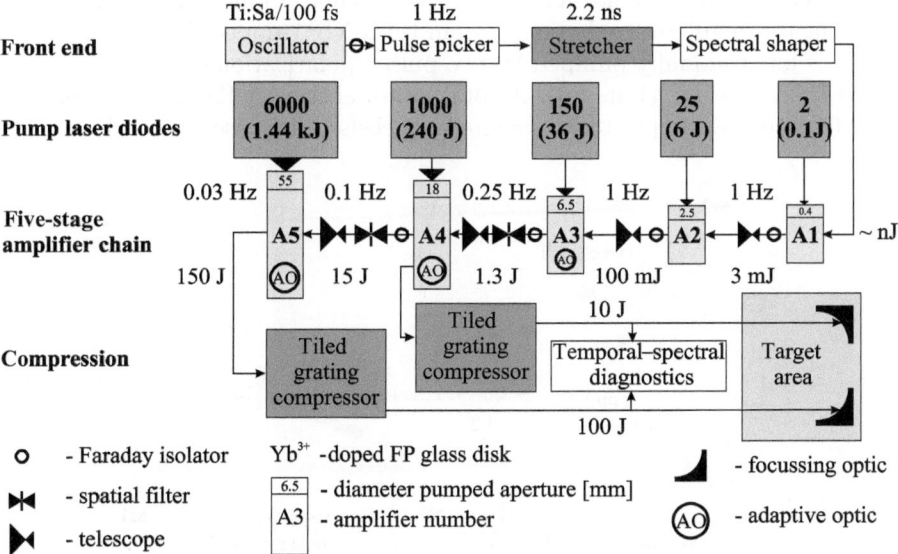

Fig. 4.5. The outline of the POLARIS system. Pulses as short as 100 fs are generated in a commercial oscillator, stretched, amplified in five stages labelled A1 to A5 and recompressed at the 100 TW level with a first tiled grating compressor and at full power by a second one

point distance of two times the desired bandwidth of the laser system ensures an acceptable pulse contrast ratio. For convenient control of the pulse chirp a dazzler [23] is inserted between the stretcher and the first amplifier. The stretched pulse energy of less than 0.2 nJ is boosted to the 150 J-level by five diode-pumped amplification stages labelled A1 to A5 (Fig. 4.5).

The amplifiers A1 and A2 are designed as regenerative amplifiers. At their energy level and corresponding beam size a stable cavity can be found whereas this is no longer possible for the successive stages. The multipass configurations incorporate angular as well as polarization multiplexing for different passes. In all amplifiers the pump as well as the final laser pulse fluence is kept constant allowing for a maximum energy extraction. A small compressor after amplifier A4 is able to generate pulses having a peak power in the 100 TW range. Once the technology of tiled grating compression has progressed sufficiently a second compressor chamber for full power will be added to the system.

The target chamber will be located in a separate subterrestrial target area, wich provides enough radiation shielding for high-intensity laser target interaction experiments.

4.5 The Five Amplification Stages of POLARIS

4.5.1 The Two Regenerative Amplifiers A1 and A2

The regenerative amplifier A1 increases the oscillator pulse energy from ≤ 1 nJ by a factor of 10^7 to 3 mJ. The 6.5-mm thick Yb^{3+}-doped fluoride phosphate glass is longitudinally pumped by two pulsed polarization coupled laser diode bars from one side (Fig. 4.6). A total pump energy of 120 mJ is available in a 2.6-ms pulse. A quarter wave single pockels cell is used for switching the

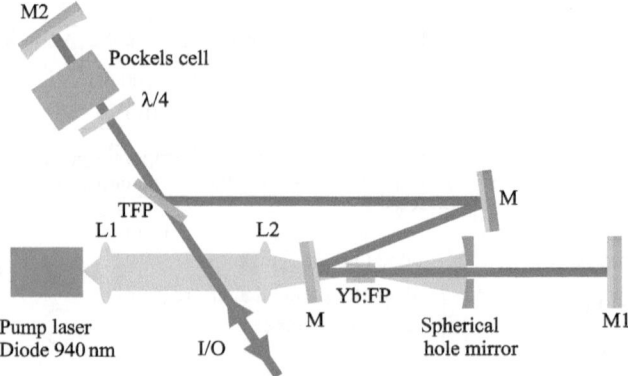

Fig. 4.6. Schematic of the regenerative amplifier A1, capable of amplifying pulses of less than a nJ energy to more than 3 mJ with a bandwidth of 13 nm

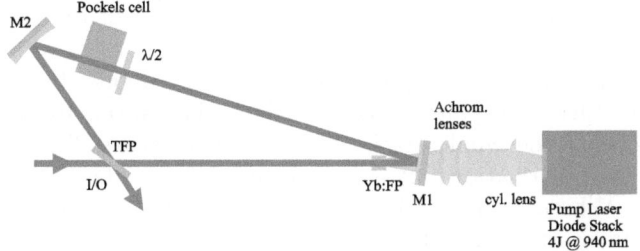

Fig. 4.7. The regenerative amplifier A2, pumped by a 25 bar diode stack

pulse into and out of the cavity where it remains for about 100 round trips. Because of its amplification factor of 10^7, this amplifier introduces the main center wavelength shift that is common to quasi-three-level laser amplifiers and also most of the gain narrowing. The bandwidth is reduced to 13 nm and the center wavelength shifted to 1032 nm. Both parameters do not change remarkably in the subsequent amplification stages.

The next level of amplification is achieved by the second regenerative amplifier A2 shown in Fig. 4.7. The cavity is designed to support a mode waist of 1.8 mm. The beam waist of the pure cavity is located in the glass to minimize the influence of thermal effects. The ring cavity consists of a dichroic mirror transmitting the pump light, a polarizer in reflection, and a spherical mirror to achieve a stable cavity. The amplifier is pumped by a laser diode stack consisting of 25 bars. It delivers an output energy of 5 J in a 2.6-ms pulse. To drive the laser diode stacks we are using pulsed diode drivers, allowing stabilized rectangular pulses of up to 250 A and 60 V with rise and fall times of 80 μs. Output energies of 100 mJ were achieved with amplifier A2.

4.5.2 The Multipass Amplifiers A3 and A4

Figure 4.8 shows the setup of the amplifier A3. By keeping the energy densities the same as in A2, the laser beam diameter is further increased to amplify

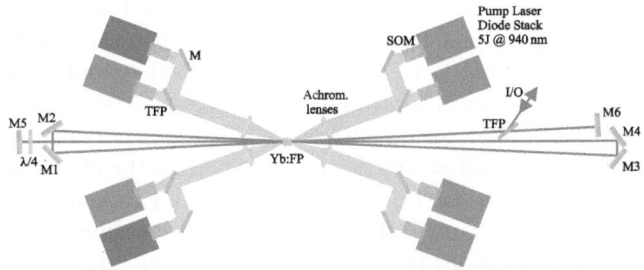

Fig. 4.8. Setup of the third amplifier A3 of the POLARIS laser system with a maximum stable output pulse energy of 1.25 J. With polarization rotation after two times 3 passes, 12 passes could be achieved in a compact design

pulses to the Joule level. A glass disc with a diameter of 12 mm and a length of 13 mm is mounted to a water-cooled heat sink.

The multipass amplifier is pumped by eight laser diode stacks as described before. The stacks are pairwise coupled by polarization. In order to rotate the polarization, the whole stack was rotated in space. This does not only influences the polarization but also reverses slow and fast axis of the diode laser beam. Whereas the focal points result in elliptical beam profiles with horizontal and vertical orientations, the pump beams are adjusted to achieve a homogeneous square like pump distribution.

The beam path is shown in Fig. 4.8. After passing a telescope, a thin-film polarizer (TFP), a Faraday rotator, and a half-wave plate which are not shown in Fig. 4.8, another TFP reflects it into the multipass beam path where it passes the amplifying glass six times. Mirror M5 reflects the pulse which is now rotated in polarization by a double pass through the quarter wave plate in front of M5 back to the TFP. This polarizer now transmits the beam which is again back reflected by M6 for another six passes through the laser medium. After passing the quarter wave plate twice again its polarization then matches its originally orientation and the amplified pulse exits into the direction of the seed after a total number of 12 passes. Both beams are separated by a Faraday rotator combined with a second TFP which are not shown here.

Stable pulses with energies of 1.25 J have been measured with an absorbed pump energy of 25 J, which corresponds to a diode current of 150 A. In this case the seed pulse energy from amplifier A2 was 80 mJ and the repetition rate 0.2 Hz [24]. By increasing the pump energy to 35 J, a maximum output of 2 J was achieved. The output energy is limited by the damage threshold of about 3 J/cm^2. For daily operation the output energy is set to 1.5 J.

In the multipass laser amplifier A4 an energy of 240 J of light at 940 nm is used to pump a 13-mm-thick laser glass disk with a diameter of 28 mm. Figure 4.9 shows an outline of the pump arrangement with the diode stacks, the collimating and directing optics, the focusing optics assembled as two lens rings and the laser glass disk. The radiation of 40 diode stacks each consisting of 25 laser diode bars is focussed onto the circular area of the disk. The pump area has a diameter of 18 mm. Forty directing mirrors and 160 cylindrical collimating and focusing lenses of coated fused silica deliver the pump light out of the net area of the emitting diode stacks of 200 cm^2 to the 80 times smaller pump area. A two-sided ring-shaped assembly of diode stacks and attached optics gives an optimal pumping geometry in terms of packing density, acceptance angle, and path length [25].

Because of the imaging properties for the fast and the slow axis the focus of a laser diode stack can be described as an elliptical, gaussian-shaped intensity profile. Because of mechanical and geometrical limitations the optical path length from the laser diodes to the amplifier medium cannot be shorter than 800 mm. The acceptance angle is 21°. Concerning the low beam quality of the laser diode's slow axis, two cylindrical lenses are used for collimation and a cylindrical lens for focusing in a distance of 200 mm to the glass. The fast axis

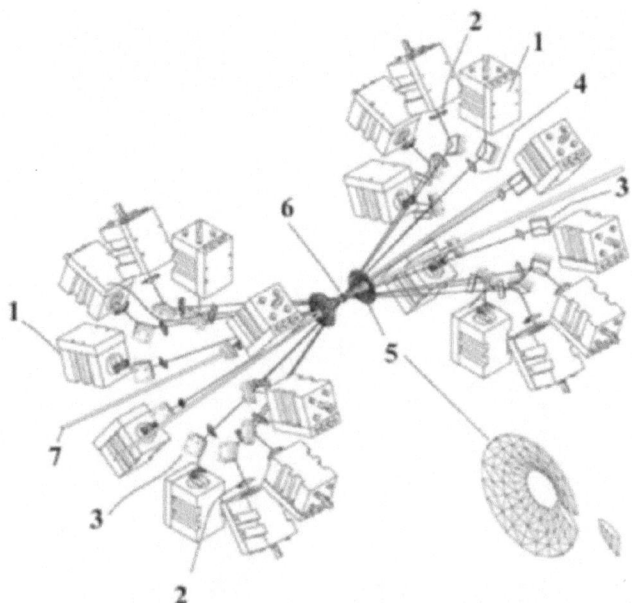

Fig. 4.9. Pump arrangement of POLARIS laser amplifier A4: **1** stack with 25 laser diode bars, **2** slow axis collimation and fast axis focusing lenses, **3** adjustable directing mirror with special HR coating, **4** slow axis collimation lens, **5** slow axis focusing ring lens system, **6** Yb^{3+}-doped fluoride phosphate glass, **7** multipassed laser pulse

is directly focussed into the glass by a thin cylindrical lens with a focal length of 800 mm. On average, the focus dimensions are measured to be 4 mm for the slow axis and 8 mm for the fast axis, respectively.

To achieve a smooth top-hat-shaped sum profile, 2×20 single pump light spots have to be spread in even balance over the required area of $2 \times 2.5\,cm^2$ by adjusting the 40 directing mirrors.

So far 8 J of output pulse energy have been achieved with this amplifier. The pulses exhibit a bandwidth of 12 nm. In near future, an operation with pulses having 15–20 J is expected while the construction of the last amplifier A5 is going on.

4.5.3 A Design for the Amplifier A5

The final amplification of the pulses from 15 J out of multipass amplifier A4 to at least 150 J is provided by multipass amplifier A5. To ensue an amplification below the damage fluence with comparable gain as in laser amplifier A4 a minimum pump energy of 1.4 kJ is required to be focussed onto the laser medium. The 13-mm-long Ytterbium fluoride phosphate glass disk is 70 mm in diameter, whereas the pumped region is $25\,cm^2$.

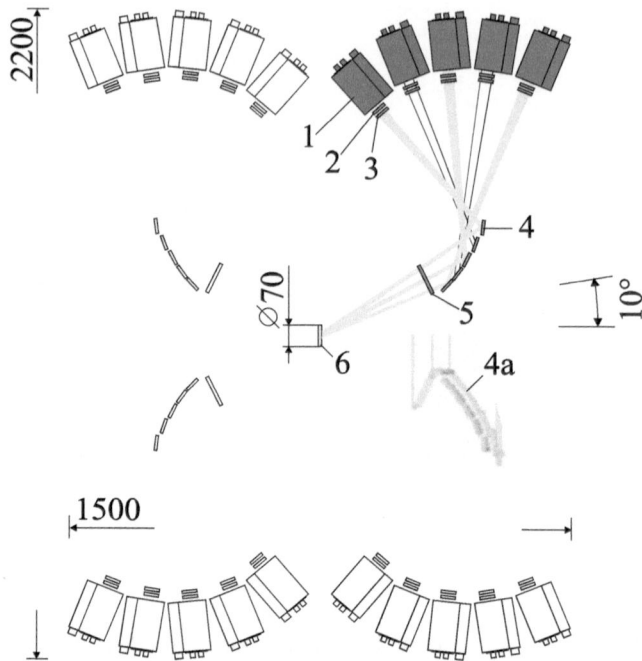

Fig. 4.10. Pump diode stack arrangement of POLARIS laser amplifier A5, 2×24 modules with five laser diode stacks, five collimating lenses, directing mirrors and one slow axis focusing lens. Lengths are given in mm

Two hundred forty laser diode stacks each delivering 6 J at 940 nm are arranged in a ring-shaped assembly comparable to A4. Applying lens optics the entire pump radiation out of a net area of 2500 cm^2 is imaged onto the 50 times smaller gain medium. In order to minimize the acceptance angle for the pump light, five laser diode stacks are grouped in a module consisting of five adjustable mirrors and a final slow axis focusing lens. Figure 4.10 shows the setup of the pump arrangement of laser amplifier A5. The pump light is designed to be p-polarized with respect to the directing mirrors. Because of limited beam quality of the laser diodes along their slow axis, additional collimation is necessary close to the laser diode stacks. The fast axis component of the pump radiation is focussed with a 700-mm lens, whereas the entire optical path length is 1200 mm. Moving the fast axis focal point close to the directing mirror and the slow axis focusing lens results in small optical components. Elliptically shaped pump light spots with an average size of 6×10 mm^2 are imaged onto the laser glass.

The resulting pump illumination on the laser glass is homogenized by individually positioning the pump light spot of each laser diode stack similar to the setup of amplifier A4. Four hundred eighty stepper motors controlled by a computer algorithm [26] are placed behind the directing mirrors to adjust one axis each.

Similar to laser amplifier A4, the A5 resonator is a multipass cavity. At the present stage a prototype of one pump module that consists of five laser diode stacks and the attached optics is accomplished.

4.6 The Tiled Grating Compressor

The development of laser systems based on the technique of chirped-pulse amplification (CPA) made it possible to generate pulses which reach focused intensities in excess of 10^{21} W/cm^2. The most crucial part of every CPA laser system is the grating compressor. Its efficiency should be as high as possible. Moreover, system performance is often limited by the damage threshold of the compressor gratings. To generate pulses in the PW-regime these pulses must be stretched temporally by a factor of 10^4 or more before amplification in order to prevent laser-induced damage in the laser material. The recompression of such extremely chirped pulses in a Treacy compressor design [27] leads to a grating separation in the range of several meters. A lateral beam size in the range of 1 m on the surface of the second grating is the consequence.

In this section we investigate the setup of a folded two-grating compressor containing two gratings and a highly reflective end mirror. The grating used for the first and last pass must be designed to withstand high fluences, since the product of ultrashort laser pulse damage threshold and usable grating area restricts the maximum energy throughput. The second grating (passes no. 2 and 3) is optimized for efficiency and diffracting area. However, since meter-sized gratings are hardly available at present, the only alternative for high-power CPA lasers is the use of tiled gratings for the second and third grating pass in compressor.

When adjusting one grating tile to another, 5 degrees of freedom have to be taken into account as shown in Fig. 4.11. In case of perfectly aligned gratings the angle of excidence from the second grating equals the angle of incidence on the first grating for all wavelengths. Since no imaging optics are involved, collimated wave packets remain collimated after the passage through the compressor. However, this is not the case, if the compressor grating pairs are not strictly parallely aligned [28], since misalignment with respect to all three rotational degrees of freedom leads to an angular separation of the beamlets in the far field, each being formed by one half of the beam passing one of the two parts of the tiled grating. Their separability depends on the laser wavelength and the diameter of the beam. For the 100 TW compressor following A4, a beam diameter of 120 mm, a central wavelength of 1030 nm and assuming $M^2 = 1$ the focal images become separated for an angular misalignment of more than 10 µrad.

From theory it was derived that a rotational misalignment of 1 µrad of the mosaic grating with respect to the reference grating leads to an angular beam misalignment of 2.5 µrad (twist, see Fig. 4.11), 3.3 µrad (tip), and 5.0 µrad (tilt) after the compressor, respectively. Hence, the accuracy of alignment of

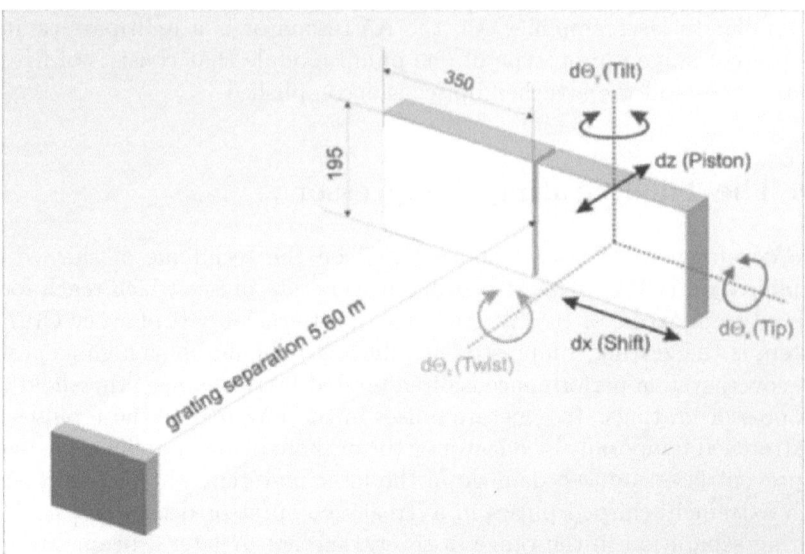

Fig. 4.11. Setup: A pair of gratings in a tiled grating compressor system. There are 5 degrees of freedom (2 translational and 3 rotational) for each mosaic grating with respect to the reference grating (in this figure the left grating is set to be the reference)

the rotational mechanical stages must be better than a few μrad. In this case the optical retardation is in the range of a few wavelengths only, leading to a inhomogeneous temporal broadening of less than 2% of the pulse width [29]. Thus, the acceptable spatial broadening of the focal spot due to angular misalignment defines the necessary alignment accuracy. The mechanics used to provide such a degree of accuracy is shown in Fig. 4.12.

Misalignment in the translational degrees of freedom (shift, piston; refer Fig. 4.11) leads to a phase difference of the wavefronts of the beamlets stemming from the different gratings. When this phase shift is not equal to zero, the beam focus is split into two parts [30]. Concerning shift, the grating grooves at the adjacent edges of the grating tiles should be separated by a multiple of the grating constant not to produce a phase shift, while for the piston the optical path difference must be a multiple of the wavelength.

The measurement setup used to detect all possible kinds of mosaic grating misalignment is shown in Fig. 4.13. For the detection of rotational misalignment the focusing of the zeroth-order reflection of a cw-laser beam with largest possible diameter and lowest possible wavelength is sufficient. For the detection of piston, two zeroth-order reflections from different angles are necessary. Only if both reflections show no phase shift, the grating surfaces are in one plane. A phase difference due to a misalignment in shift can only be detected in a diffraction order unequal to zero. Thus, two zeroth-order reflections and one minus first-order reflection are needed to adjust the mosaic grating tiles in a phase-true manner.

Fig. 4.12. Mosaic Grating setup used within the POLARIS system at IOQ Jena

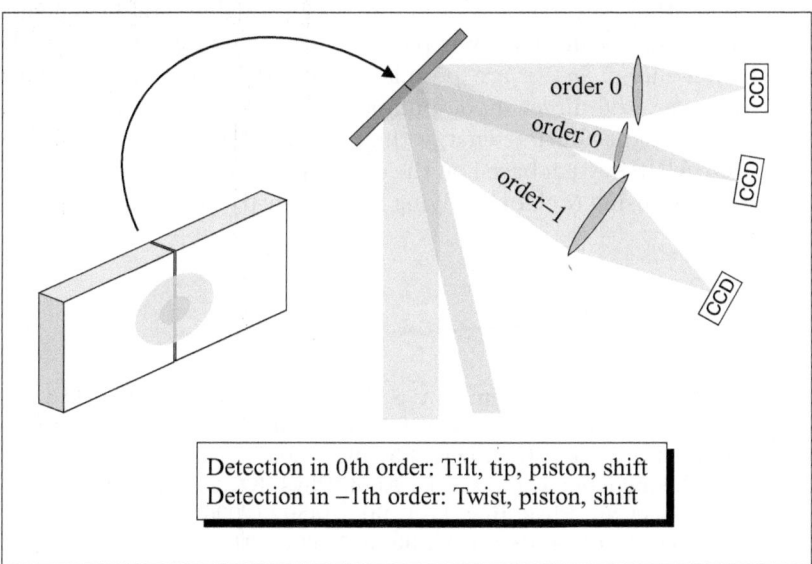

Fig. 4.13. Setup used for detecting misalignment errors of the mosaic gratings with respect to the reference grating

4.7 Future Prospects

The POLARIS project has already shown that an all diode-pumped amplification of 150-fs pulses to the 10 J level is possible. Although, the amplifying medium is a glass with its typical small thermal conductivity a repetition rate much higher than in existing flashlamp pumped lasers of the same energy class can be reached. Therefore, diode pumping offers the opportunity for significantly higher repetition rates in high power CPA laser systems [31] than using conventional flash lamp-pumped Nd:glass laser systems.

The application of advanced cooling techniques allows a further improvement of the repetition rate of CPA lasers. The feasibility of these techniques has already been successfully demonstrated in high average power diode-pumped laser systems like MERCURY [32], HALNA [33, 34], or LUCIA [35, 36].

In addition, broad bandwidth laser crystals doped with Yb^{3+}, for example, $Yb:CaF_2$ [37, 38], Yb:BOYS [39], Yb:LSO, and Yb:YSO [40, 41], or Yb:KGW and Yb:KYW [42] can improve heat removal from the amplifier at comparable conditions.

An alternative to CPA systems for the generation of very short pulses with high energies is optical parametric chirped pulse amplification (OPCPA) [43, 44, 45, 46, 47]. Existing projects to generate PW-pulses [48, 49] emphasize the OPCPA potential. Nevertheless, a pump laser providing stable synchronized pulses with a high beam quality is required for the nonlinear process. Diode-pumped high-energy lasers are promising candidates for an efficient pump source. However, the direct diode-pumped amplification of ultrashort pulses can result in a higher output energy than the indirect way of OPCPA.

The POLARIS system based on the direct amplification architecture represents a very promising tool for studying high-intensity laser–matter interaction physics.

References

1. D. Strickland and G. Mourou: Optics Communications **56**, 219 (1985)
2. E. L. Clark et al.: Phys. Rev. Lett. **85**, 1654 (2000)
3. R. Snavely et al.: Phys. Rev. Lett. **85**, 2945 (2000)
4. K. W. D. Ledingham: J. Physics D: Appl. Phys. **37**, 2341 (2004)
5. H. Schwoerer et al.: Phys. Rev. Lett. **96**, 014802 (2006)
6. M. Kaluza et al.: Phys. Rev. Lett. **93**, 045003 (2004)
7. C. B. Edwards et al.: Central Laser Facility, Annual Report, pg. 164 (2001)
8. Y. Kitagawa et al.: IEEE J. Quantum Electron. **40**, 281 (2004)
9. M. Aoyama et al.: Opt. Lett. **28**, 1594 (2003)
10. C. Palmer: *Diffraction Grating Handbook* (Richardson Grating Laboratory, 2002)
11. W. Koechner: *Solid State Laser Engineering* (Springer, 1998)
12. M. Siebold et al.: to be published

13. D. Ehrt: Curr. Opin. in Solid State Mater. Sci. **7**(2), 135 (2003)
14. D. Ehrt, T. Töpfer: In: Proc. SPIE Int. Soc. Optical Engineering vol. **4102**, 95 (2000)
15. T. Danger, E. Mix, E. Heumann, G. Huber, D. Ehrt, W. Seeber: OSA Trends in Optics and Photonics on Advanced Solid State Lasers pp. 23–25 (1996)
16. E. Mix, E. Heumann, G. Huber, D. Ehrt, W. Seeber: Advanced Solid-State Lasers Topical Meeting Proc. **1995**, 230 (1995)
17. V. Petrov, U. Griebner, D. Ehrt, W. Seeber: Opt. Lett. **22**(6), 408 (1997)
18. S. Paoloni, J. Hein, T. Töpfer, H.G. Walther, R. Sauerbrey, D. Ehrt, W. Wintzer: Appl. Phys. B **78**, 415 (2004)
19. J. Hein, S. Paoloni, H.G. Walther: J. Phys. IV France **125**, 141 (2005)
20. T. Töpfer, J. Hein, J. Philipps, D. Ehrt, R. Sauerbrey: Appl. Phys. Lasers Opt. **B71**(2), 203 (2000)
21. F. Dorsch, V. Blümel, M. Schröder, D. Lorenzen, P. Hennig, D. Wolff: Proc. SPIE **3945**, 43 (2000)
22. R. Göring, S. Heinemann, M. Nickel, P. Schreiber, U. Röllig: german patent No. DE 198 00 590 A1, Jenoptik AG Jena (1999)
23. F. Verluise, V. Laude, Z. Cheng, C. Spielmann, P. Tournois: Opt. Lett. **25**(8), 575 (2000)
24. J. Hein, S. Podleska, M. Siebold, M. Hellwing, R. Bödefeld, R. Sauerbrey, D. Ehrt, W. Wintzer: Appl. Phys. B **79**, 419 (2004)
25. J. Philipps, J. Hein, R. Sauerbrey: patent No. DE 102 35 713 A1 (2004)
26. M. Siebold, S. Podleska, J. Hein, M. Hornung, R. Bödefeld, M. Schnepp, R. Sauerbrey: Appl. Phys. B (in press)
27. E.B. Treacy: IEEE J. Quantum Electron. **5**, 454 (1969)
28. G. Pretzler, A. Kasper, K.J. Witte: Appl. Phys. B **70**, 1 (2000)
29. C. Fiorini, C. Sauteret, C. Rouyer, N. Blanchot, S. Seznec, A. Migus: IEEE J. Quantum Electron. **30**, 1662 (1994)
30. T. Kessler, J. Bunkenburg, H. Huang, A. Koslov, D. Meyerhofer: Opt. Lett. **29**, 635 (2004)
31. C.P.J. Barty: Techn. Digest (1998)
32. J.T. Early: AIP Conf. Proc. **(578)**, 713 (2001)
33. T. Kawashima, T. Kanabe, H. Matsui, T. Eguchi, M. Yamanaka, Y. Kato, M. Nakatsuka, Y. Izawai, S. Nakai, T. Kanzaki, H. Kan: Jpn. J. Appl. Phys. Pt.1 Regular Papers, Short Notes & Rev. Papers **40**(11), 6415 (2001)
34. S. Nakai, T. Kanabe, T. Kawashima, M. Yamanaka, Y. Izawa, M. Nakatuka, R. Kandasamy, H. Kan, T. Hiruma, M. Niino: Proc. SPIE Int. Soc. Opt. Eng. **4065**, 29 (2000)
35. J.C. Chanteloup, G. Bourdet, A. Migus: Technical Digest CLEO No. 02CH37337 pp. 515–516 (2002)
36. J.C. Chanteloup, G. Bourdet, A. Migus: Technical Digest CLEO No.02CH37337 (2002)
37. A. Lucca, G. Debourg, M. Jacquemet, F. Druon, F. Balembois, P. Georges: Opt. Lett. **29**(23), 2767 (2004)
38. V. Petit, J. Doualan, P. Camy, V. Ménard, R. Moncorgé: Appl. Phys. B **78**, 681 (2004)
39. F. Druon, S. Chénais, P. Raybaut, F. Balembois, P. Georges, R. Gaumé, G. Aka, B. Viana, S. Mohr, D. Kopf: Opt. Lett. **27**(3), 197 (2002)
40. M. Jacquemet, C. Jacquemet, N. Janel, F. Druon, F. Balembois, P. Georges, J. Petit, B. Viana, D. Vivien: Appl. Phys. B **80**, 171 (2005)

41. F. Druon, S. Chénais, F. Balembois, P. Georges, R. Gaumé, B. Viana: Opt. Lett. **30**(8), 857 (2005)
42. G. Paunescu, J. Hein, R. Sauerbrey: Appl. Phys. B **79**, 555 (2004)
43. P. Matousek, I.N. Ross, J.L. Collier, B. Rus: Technical Digest CLEO No.02CH37337 p. 51 (2002)
44. G. Cerullo, S. De-Silvestri: Rev. Sci. Instrum. **74**(1), 1 (2003)
45. P. Matousek, B. Rus, I. Ross: IEEE J. Quantum Electron. **36(2)**, 158 (2000)
46. P. Matousek, I.N. Ross, J.L. Collier, B. Rus: Techn. Digest (2002)
47. I.N. Ross, J.L. Collier, P. Matousek, C.N. Danson, D. Neely, R.M. Allott, D.A. Pepler, C. Hernandez-Gomez, K. Osvay: Appl. Opt. **39**(15), 2422 (2000)
48. Zhu-Peng-Fei, Qian-Lie-Jia, Xue-Shao-Lin, Lin-Zun-Qi: Acta Physica Sinica **52**(3), 587 (2003)
49. M. Nakatsuka, H. Yoshida, Y. Fujimoto, K. Fujioka, H. Fujita: J. Korean Phys. Soc. **43**(4), pt. 2, 607 (2003)

5

The Megajoule Laser – A High-Energy-Density Physics Facility

D. Besnard

CEA/DAM Île de France, BP12 91680 Bruyères le Châtel, France
didier.besnard@cea.fr

Abstract. The French Commissariat à l'Energie Atomique (CEA) is currently building the Laser Megajoule (LMJ), a 240-beam laser facility, at the CEA Laboratory CESTA near Bordeaux. The LMJ will be a cornerstone of the CEA's "Programme Simulation," the French Stockpile Stewardship Program. It is designed to deliver 1.8 MJ of 0.35 μm light to targets for high-energy-density physics experiments, among which fusion experiments. LMJ technological choices were validated with the Ligne d'Intégration Laser (LIL), a scale 1 prototype of one LMJ bundle built at CEA/CESTA. It delivered 9.5 kJ of UV light (0.35 μm) in less than 9 ns from a single laser beam in May 2003. This chapter will present results from the commissioning phase of the LIL program in 2003 and 2004. The construction of the LMJ facility itself started in March 2003. The LMJ will be commissioned early 2011, and the first fusion experiments begin late 2012.

5.1 LMJ Description and Characteristics

The LMJ facility is a key part of the French Stockpile Stewardship Program, the "Programme Simulation." The LMJ is devoted to laboratory experiments on the behavior of materials under very high temperature and pressure conditions. It has applications in the field of astrophysics, Inertial Fusion Energy (IFE), and fundamental physics [1]. It is also a key facility for training physicists engaged in the French deterrent. In order to cover these different applications, the facility is designed with the maximum flexibility in terms of pulse duration (from 200 ps to 25 ns) and power. Plasma diagnostics will be easily interchanged depending on the type of experiments and special diagnostic inserters and positioners will be common to those at the NIF facility (Lawrence Livermore National Laboratory [LLNL]) [2].

5.1.1 LMJ Performances

The Megajoule most stringent specifications are dictated by fusion experiments, the first of which to be realized late 2012. Specifications for the laser

D. Besnard: *The Megajoule Laser – A High-Energy-Density Physics Facility*, Lect. Notes Phys.
694, 67–77 (2006)
www.springerlink.com   © Springer-Verlag Berlin Heidelberg and European Communities 2006

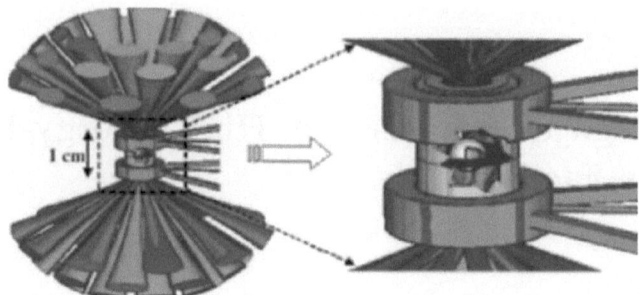

Fig. 5.1. A CAD view of the LMJ cryogenic target

were obtained with an optimization of the laser itself together with the fusion target (Fig. 5.1). A fusion target is composed of a 1-cm-long, usually gold, cylindrical hohlraum, used to convert laser light to X-rays. X-rays smoothly irradiate a 2-mm-diameter capsule, composed of a polymere ablator and DT. The laser beams enter the hohlraum through two apertures, one on each side of the hohlraum. The laser beams illuminate the capsule as quadruplets. The laser light spots locations are chosen in such a way that the converted X-rays provide a very uniform irradiation of the capsule. In order to further enhance the efficiency of the implosion, the DT contained in the capsule is solid for the most part. The fusion capsule is therefore composed of the ablator, a layer of solid DT, and a central gaseous DT core. The target assembly is maintained at about 18K with an elaborate cryogenic system. Figure 5.1 shows the high-purity aluminum target holder that refrigerates the target through thermal conduction.

To determine LMJ specifications, we used numerical simulation to optimize 1D capsules imploding under shaped laser pulses. This gave required energy and laser power. Two-dimensional (2D) integrated simulations were then performed to optimize beams position and energy balance. Such calculations accounted for laser plasma interaction within the target's hohlraum, as well as symmetry requirements. Margins were added to the reference design to take into account the two main remaining uncertainties, that is, the effect of parametric and hydrodynamic instabilities.

With 240 beams arranged in 30 bundles (8 beams per bundle), LMJ will deliver 1.8 MJ of UV light (0.35 m). For pulse durations of about 3.5 ns, the corresponding power will be about 550 TW and significant target gains are expected with cryogenic targets with indirect drive (Fig. 5.2). With these specifications, our baseline capsule design is proved to be robust. An extensive analysis was performed, which gave an estimate of the effect of 19 coupled parameters; these involve beam pointing accuracy, beam energy, as well as target dimensions and fabrication. To do so, a set of three models was adjusted to our reference 2D simulations: a raytracing model, which gives the laser flux on the hohlraum's walls, a view factor model, which gives the X-ray flux on

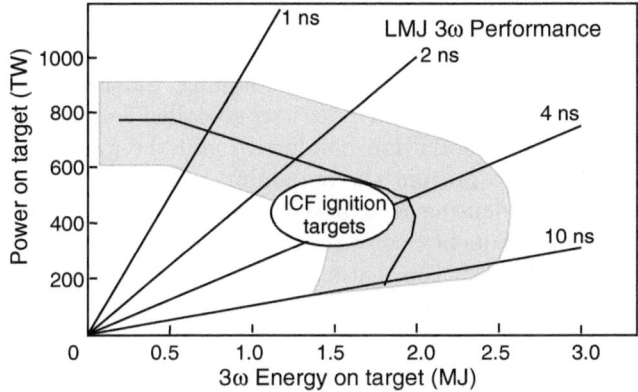

Fig. 5.2. Energy/power operating region of LMJ

the capsule, converted from laser light, and an implosion model, giving the DT final radius. Our baseline design was to be declared robust if the hot spot (within which fusion reactions occur) deformation stayed much smaller than the hot spot size. Indeed, the hot spot deformation's amplitude is always smaller than 10–15% of the hot spot size.

The facility will allow up to 400 shots per year, among them half being physics shots, including high-yield shots [3].

5.1.2 LIL/LMJ Facility Description

The LMJ has been described in previous chapters [4, 5]. It has a multipass amplification structure (Fig. 5.3), the 18 amplifier laser glass slabs being arranged in two amplifiers within a four-pass cavity whose end-cavity mirror is a deformable mirror in order to correct wavefront distortion. The front-end pulse

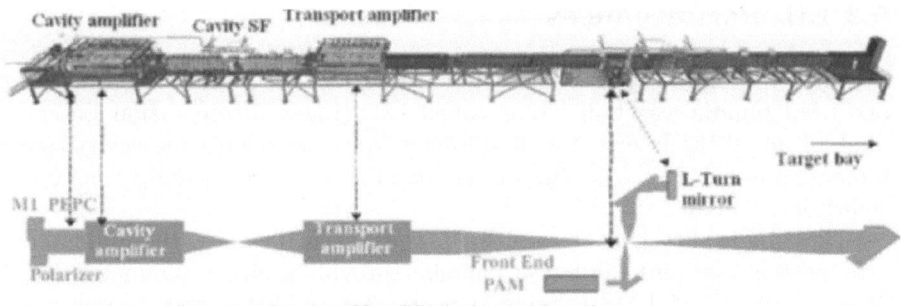

Fig. 5.3. General laser layout showing the pulse injection, the L-Turn providing a passive method for a four-pass cavity operation, the large Pockels-Cell (PEPC) for isolation and the deformable end-cavity mirror M1

(up to 1 J) is injected in the transport spatial filter and the four passes are obtained with a passive optical arrangement called "Demi-Tour" (L-turn).

In order to optimize the laser–target coupling, different beam smoothing techniques will be used, starting with the so-called longitudinal-SSD. It is naturally provided by the 0.5-nm bandwidth and the gratings used to focus the beam on target. To prevent the remaining 1ω and 2ω light from entering the target chamber, focusing is achieved by using a pair of gratings, one on each side of the two frequency conversion crystals. The second grating deflects and focuses the 3ω light at the center of the target chamber, letting the other wavelengths to be absorbed outside of the chamber (Fig. 5.4).

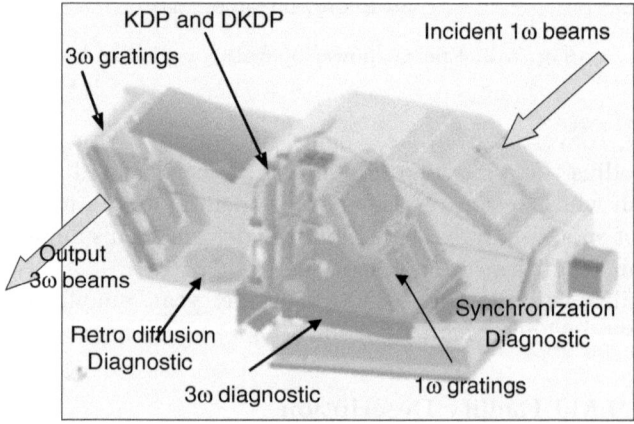

Fig. 5.4. LMJ focusing system uses two diffraction gratings to filter unconverted 1ω and 2ω light. The 2nd grating working at $0.35\,\mu\mathrm{m}$ focuses the beam on target

5.2 LIL Performances

In order to validate technological choices for LMJ, a scale 1 prototype of one LMJ bundle was built. It is called LIL (Ligne d'Intégration Laser). A photograph of the laser bay and amplification section with the cavity spatial filter is shown in Fig. 5.5. The target chamber and laser beamlines setup are shown in Fig. 5.6.

The LIL facility will be used for plasma experiments with various beam arrangements around the target chamber providing either symmetric irradiation or two-sided LMJ type bipolar irradiation. LIL's first beamline was activated in two successive phases (1ω amplifier section and 3ω conversion).

Fig. 5.5. Picture of LIL laser bay

Fig. 5.6. CAD view of the LIL beams around the target chamber

First Phase: 1ω Performance

The 1ω power experiments were conducted with a laser pulse injected from the PAM (preamplication module) through the four-pass main amplifier chain. The main amplifier output energy was ramped up from several hundred Joules to 1.8 kJ, using a 700-ps pulse (Fig. 5.7) and generating a peak power of 4 TW. The measurements were made at the output of the transport spatial filter lens, using the 1 laser diagnostic module installed on each beam.

Energy experiments were then undertaken with longer pulses up to 20 kJ at 4.2 ns in 2003 [6].

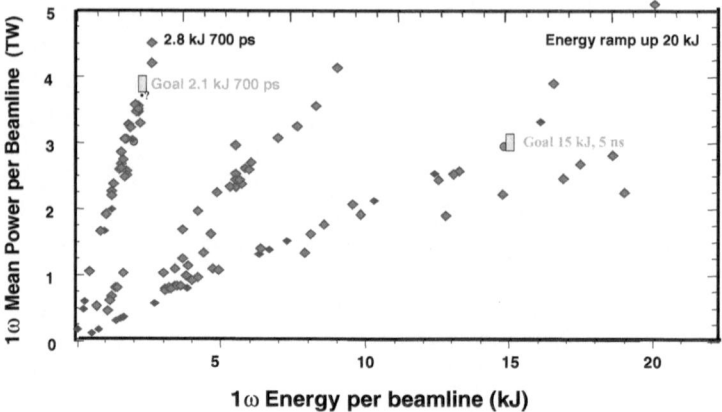

Fig. 5.7. Operating domain at 1ω

Second Phase: 3ω Performance on Target at the Center of the Chamber

These measurements took into account energy, pulse duration and temporal shape, contrast ratio, and size and profile of the focal spot. Experiments allowed the 3ω energy to be ramped up from several hundred joules to 1.5 kJ, using a 700-ps duration pulse (Fig. 5.8).

For energy ramp-up the output energy was increased from 1 kJ to 7.5 kJ with a 5-ns pulse and to 9.5 kJ with an 8.8-ns pulse. LIL was the first laser to produce 9.5 kJ of UV light in less than 9 ns in one beam.

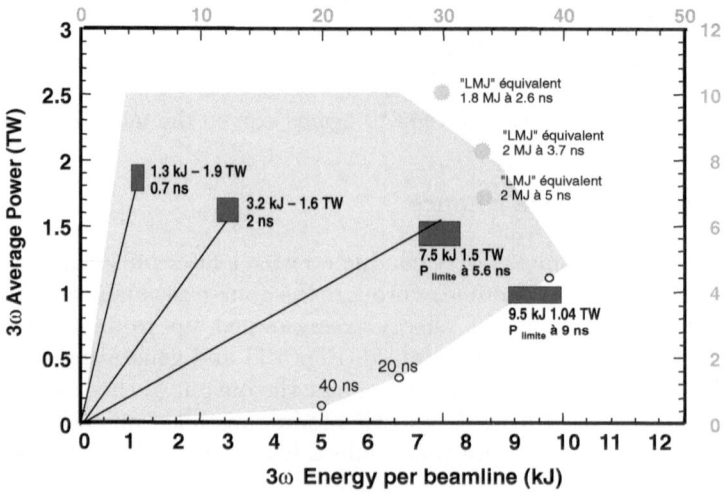

Fig. 5.8. Operating domain at 3ω

The first focal spots were measured at 1 kJ of 3ω giving a size of 500 μm (at 3% of maximum intensity) in good agreement with time integrated X-ray image (dimension of X-ray pinhole record <500 μm).

Third Phase: Quadruplet Performance

We began in 2004 the full operation of the quadruplet, and the complementary performance characterization was finished during summer 2004: beam synchronization has been demonstrated with an accuracy better than 30 ps for the four beams, while far field imaging at target chamber center is now undertaken at low intensity level. The full quadruplet performance was demonstrated with the accurate superposition of the four focal spots.

With these results, LMJ technological choices are confirmed.

5.3 LMJ Facility

The construction of the LMJ building started in 2003 (Fig. 5.9). Located 150 m away from the existing LIL building, the four laser bays will be located on both sides of the $40 \times 40\,\mathrm{m}^2$ target chamber bay.

Fig. 5.9. Artist view of the future LMJ facility

The design of the LMJ building takes into account constraints related to the target bay (Fig. 5.10), the plasma diagnostics, and the cryogenic target. Such constraints are important for mechanical stability, temperature control, and access for maintenance. The target chamber is in fabrication phase (Fig. 5.11) and will be installed in the targetbay by the end of 2006.

With a diameter of 10 m and a thickness of 10 cm of aluminum, it will provide 260 holes; 80 of them can be used as laser windows, giving the maximum flexibility for experimental arrangement of the 60 quadruplets.

Fig. 5.10. Model of the target bay

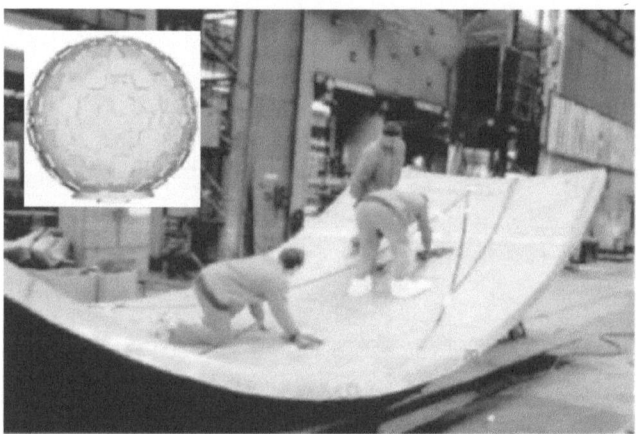

Fig. 5.11. Target chamber fabrication

Fig. 5.12. Aerial view of the construction site

The final design of the building is finished, and excavation operations are now completed (Fig. 5.12). Most of the laser and target bay concrete platforms are ready. The target chamber supporting structures construction will start in May 2005, in order to be ready to place the chamber inside the building, just before closing the roof.

The first laser bay (in the Southeast LMJ laser hall) will be ready before the end of 2006.

5.4 LMJ Ignition and HEDP Programs

LIL and LMJ experimental programs are based on a detailed analysis of CEA/DAM physics modeling needs, as identified by the Simulation Program.

LIL will first be used to test and prepare the complex experiments planned in the LMJ facility. It will also be used in its own right. Eventually, it will be coupled to a multi-kJ class petawatt laser, funded by regional and national bodies.

To achieve ignition, a comprehensive experimental program has been planned on LMJ. After a demonstration of LMJ performances, a few shots will allow to measure the effect of the smoothing techniques that have been chosen on LMJ type plasmas. Then, hohlraum experiments will be performed to adjust the radiation temperature with time. Symmetry will then be optimized, and eventually, after shock synchronization, capsule implosion will be checked against our simulations. With all these ingredients in place, fusion experiments can be performed.

LIL and LMJ are unique facilities. They will be open to the scientific community. Access to these facilities is organized by the Laser and Plasmas Institute (ILP). ILP was founded in March 2003 by the French research institutions CNRS, Bordeaux-1 University, Ecole Polytechnique, and CEA. This institute gathers about 27 laboratories in the field of high-energy lasers and high-energy-density physics.

Indeed, CEA emphasizes collaborative work in these fields. As an example, some experiments are currently proposed on LIL, to measure hydrogen's equation of state. This proposal is based on prior experiments performed on the VULCAN (2002) and OMEGA lasers [7]. Previous static experiments could induce a factor of 7 in density. Illuminating an already-compressed target with LIL beams will allow to reach an additional factor of 4. Temperatures of interest (resp. pressure) are between 0.3 and 3 eV (resp. 1 and 20 Mbar). A schematic of this experiment is shown in Fig. 5.13.

Instability modeling is also a major field of investigation. Rayleigh-Taylor instabilities at the capsule's ablation front are a source of concern when studying ignition. In a worst-case scenario, the ablator can be burned through, preventing any DT fusion. Mitigating these instabilities is therefore a crucial step on the path to ignition. Modeling of this process started some years ago,

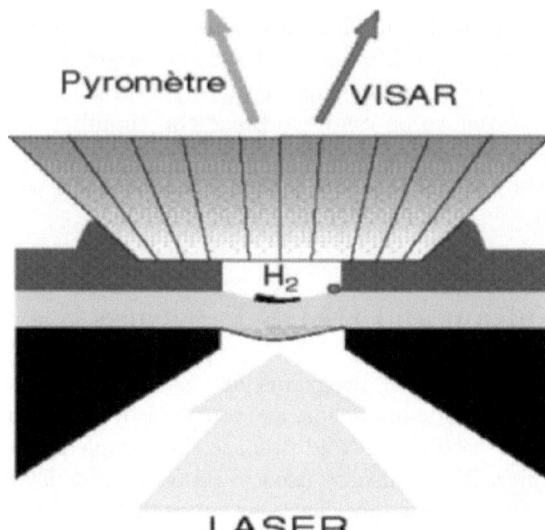

Fig. 5.13. Schematics of a laser-induced shock in a precompressed H_2 target

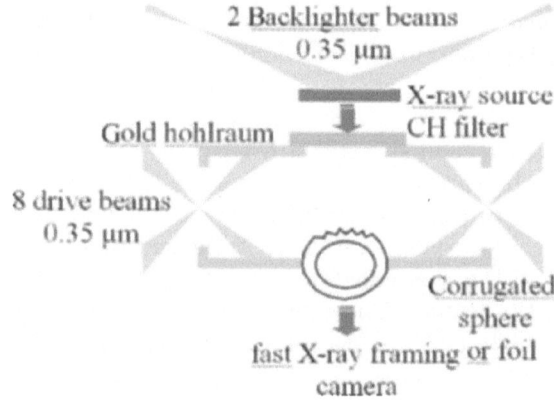

Fig. 5.14. Schematics of a laser-induced instability experiment on the NOVA laser

as well as experimental measurements at low laser energies (Fig. 5.14). Additional experiments will be performed at high energy to validate our nominal design, prior to actual fusion experiments.

5.5 Conclusions

The first step of LMJ, which consisted in the construction of a prototype, the LIL, is completed. One beamline demonstrated the goal of 20 kJ, 4 ns at 1ω and 9.5 kJ, 8.8 ns at 3ω for square pulse.

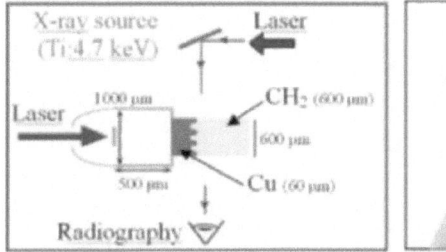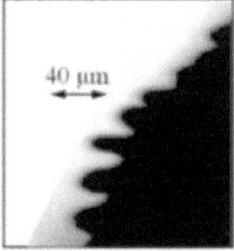

Fig. 5.15. 2.5 kJ/1.3 ns square/0.35 µm: 170 eV

The commissioning of the quadruplet was finished in September 2004. Plasma diagnostics are installed simultaneously, and they will provide the basis for the first physics experiments. The LMJ building construction is on schedule. The target chamber, which is under construction, will be introduced in the building 2006. The LMJ 240 beams will be put in operation early 2011, and fusion experiments performed late 2012.

LIL and LMJ experimental programs are in preparation. These two unique facilities will be open to the research community to address current questions in high-energy-density physics.

Acknowledgment

The author is indebted to the CEA staff for their contributions to the results presented here.

References

1. E.M. Campbell et al.: Inertial fusion science and technology for the 21st century. In: *Compte Rendus de l'Académie des Sciences*, Série IV, Tome 1, n**6** (2000)
2. E.I. Moses: The National Ignition Facility: Status and plans. In: *Proc. Current Trends in International Fusion Research: A Review*, Washington, DC, March 12–16 (2001)
3. P.A. Holstein et al.: Target design for the LMJ. In: *Proc. Inertial Fusion Sciences and Applications,* vol. 99, Bordeaux, (1999)
4. M.L. André, F. Jequier: LMJ project status. In: *Proc. Current Trends in International Fusion Research: A Review*, Washington, DC, March 12–16 (2001).
5. P. Estraillier et al.: The megajoule front end laser system overview. In: *Proc. Solid State Lasers for Application to Inertial Confinement Fusion* (ICF), Paris, October 22–25 (1996)
6. J.M. Di Nicola, J.P. Leidinger et al.: IFSA 2003 Conference
7. High Pressure Rev. **24** (2004)

Part II

Sources

6

Electron and Proton Beams Produced by Ultrashort Laser Pulses

V. Malka, J. Faure, S. Fritzler, and Y. Glinec

Laboratoire d'Optique Appliquée – ENSTA, CNRS UMR 7639, Ecole
Polytechnique Chemin de la Humiére, 91761 Palaiseau, France
malka@enstay.ensta.fr

Abstract. It is known that relativistic laser–plasma interactions can induce accelerating fields beyond 1 TV/m. Such electric fields are capable of efficiently accelerating plasma background electrons as well as protons. Depending on the target medium, high-quality particle beams can be generated. An introduction to the current state of the art will be given, and possible applications of these optically induced charged particle beams will be discussed.

6.1 Introduction

Since their discovery, beams of particles like electron and proton beams have been of great interest and relevance in various scientific domains. The evolution of the quality of these beams, by extending one of their properties (like emittance, bunch length, or energetic distribution), is always associated to new investigations and sometimes to new discovery. For example, a higher luminosity is obviously preferential for high-energy-physics experiments where the number of events for a given phenomenon is very small. Similarly, shorter particle bunches permit the investigation of phenomena with higher temporal resolution. For high-resolution radiography experiments, the electron beam should have a small, point-like source to enhance the resolution. This can be achieved with high-quality beams with low emittances. Finally, a reduction of the size of accelerators can reduce the total cost since it corresponds directly to a reduction of the cost of the infrastructure. Today, the most efficient pulsed electron sources are photo-injector guns, where lasers with energies of some tens of μJ and pulse durations of some ps irradiate cathodes and liberate electrons. However, in this case, these lasers are not intended to accelerate electrons to high energies. With the advent of the Chirped Pulse Amplification (CPA) [1], high-power, sub-ps laser pulses have become available. Focusing such lasers down to focal waists of some μm and intensities beyond 10^{18} W/cm^2, intrinsic electric fields of several TV/m can be obtained. At such high intensities, these lasers can create quasi-instantaneously plasmas

V. Malka et al.: *Electron and Proton Beams Produced by Ultrashort Laser Pulses*,
Lect. Notes Phys. **694**, 81–90 (2006)
www.springerlink.com © Springer-Verlag Berlin Heidelberg and European Communities 2006

on the targets they are focused onto; that is, they generate a medium consisting of free ions and electrons. Inside this plasma, the transverse electric laser fields can be turned into longitudinal plasma electron oscillations, known as plasma waves, which are indeed suitable for electron acceleration [2]. In addition, because of the high laser intensity, strong quasi-static electric fields can be induced, which are capable of subsequently accelerating ions.

In this chapter, we will give a brief overview on the theoretical aspects of charged particle generation induced by relativistic laser–plasma interactions. Recent experiments on electron generation as well as on proton generation will be described, and an outlook on near-future experiments will be given. Finally, possible applications of these charged particle sources will be discussed.

6.2 Theoretical Background

6.2.1 Electron Beam Generation in Underdense Plasmas

Electron beams can be generated by the breaking of relativistic plasma waves in an underdense plasma (plasma with an electron density below the critical density, $n_e < n_c$, this condition is necessary for allowing the laser to propagate in the plasma). For a laser with power exceeding few tens of TW propagating in a plasma with density values higher than a few $10^{18}\,\mathrm{cm}^{-3}$, more precisely for laser and plasma parameters satisfying $P_L/P_c >$ few unit (P_L is the laser power and the $P_c(GW) = 17n_c/n_e$, the critical power for relativistic self-focusing), relativistic plasma waves can be excited, reaching the wave breaking limit, which permits the generation of an energetic, collimated, and forward propagating electron beam. These electron beams were initially demonstrated using an energetic laser beam working in the ps range at low repetition rate (one shot/20 min) [3]. But these beams are now currently being produced with shorter laser pulses for which the required energy is considerably reduced, allowing laser plasma accelerators to work with a 10 Hz repetition rate. Using 10 Hz, 100-fs laser, electron beam with energy up to 10 MeV has been accelerated directly by the laser (DLA) [4], whereas using 10 Hz, 30 fs at lower electron density, 70 MeV energies have been reached because of an efficient acceleration by plasma waves [5]. This is illustrated on Fig. 6.1: the maximum electron energy increases when the phase velocity of plasma wave increases (i.e., when the electron density decreases).

In addition, the continuous line indicates that for densities below $10^{19}\,\mathrm{cm}^{-3}$, the expected value of the gain is well below the theoretical one. This is because of (i) very intense radial electric field and (ii) a Rayleigh length shorter than the dephasing length can limit the optimum gain. To overcome this problem, longer laser–plasma interaction lengths are needed. They can be achieved by using a lower optic aperture or by using a plasma channel. Using a long off-axis parabola, a 30-fs laser pulse propagating at very low density can excite plasma waves with amplitudes corresponding to a highly nonlinear regime of

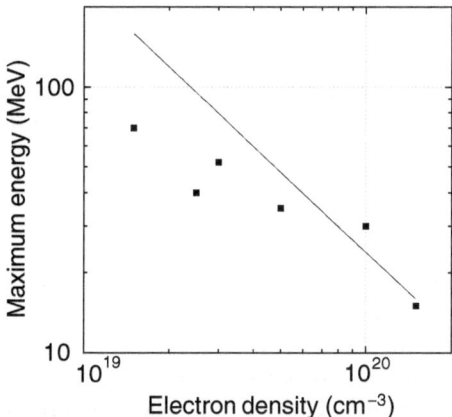

Fig. 6.1. Maximum electron energy obtained with a 30-cm off-axis parabola with a 0.6-J laser energy, 35-fs laser pulse focused down a 6-µm focal spot (laser intensity about $2 \times 10^{19}\,\mathrm{W/cm^2}$) as a function of the electron density. The theoretical value deduced from the linear theory $W_{\mathrm{max}} \approx 4\gamma_{\mathrm{p}}^2 (E_z/E_0)mc^2$, where γ_{p} is the plasma wave. Lorentz factor has been plot for $E_z/E_0 = 0.5$, where E_z/E_0 is the electrostatic field normalized to $E_0 = cm\omega_p/e$

great interest for the production of goodquality electron beams [6]. Normalized emittance was found to be as low as $(2.7 \pm 0.9)\pi$ mm mrad for (54 ± 1) MeV electrons [7]. At densities even lower, for which laser pulse length $c\tau_L$, is smaller than the plasma wavelength λ_{p}, electrons beams with prodigious parameters have recently been produced [8].

In the forced laser wakefield (FLWF) regime, occuring when $P_L > P_c$ and $c\tau_L \approx \lambda_p$, nonlinear interaction can excite nonlinear plasma waves when propagating over long enough distances. Using these conditions, a combination of laser beam self-focusing, front edge laser pulse steepening, and relativistic lengthening of the plasma wave wavelength can result in a forced growth of the wakefield plasma wave [6, 9]. Since in the FLWF regime the interaction of the bunch of accelerated electrons with the laser is reduced, this can yield the highest known electron energy gains attainable with laser–plasma interactions. Since the laser interaction with the plasma wave and with the electron beam is reduced, the generated electron beam has a very good spatial quality with an emittance as good as those obtained in conventional accelerators.

Using the same laser (pulse duration, laser energy, and focusing aperture), by carefully scanning the electron density, we observe a monoenergetic feature in the electron distribution in agreement with three-dimensional (3D) PIC simulations performed by Prof. A. Pukhov and Prof. J. Meyer-ter-Vehn, in a new "light bullet regime" [10]. During its propagation in the underdense plasma, the laser excites relativistic plasma waves, since its power exceeds the one for self-focusing, the laser radius is reduced by a factor of 3, producing a laser beam with parameters well adapted to excite resonantly a nonlinear plasma wave. At this point, the laser ponderomotive potential expels the plasma electrons radially and leaves a cavitated region (or "plasma bubble") behind. Electrons from the wall of the bubble are then injected and accelerated inside the bubble. Since they have a well-defined location in phase space, they form a high-quality electron beam.

6.2.2 Proton Beam Generation in Overdense Plasmas

In contrast, proton beams are more efficiently generated in overdense plasmas, $n_e > n_c$. Even though the laser beam cannot propagate through the overdense medium, its ponderomotive force accelerates electrons in the plasma skin layer. This force is responsible of two ions acceleration mechanisms: (i) ponderomotive and (ii) plasma sheath acceleration. The ponderomotive force expels the electrons from the high-field regions, setting up a charge imbalance that accelerates the ions in turn. This mechanism includes forward ion acceleration at the surface of an irradiated solid target [11]; it is very sensitive to the state of the surface at the front side as well as to the size of the preplasma. In the second process, plasma sheath acceleration, the forward ion beams properties are more related to the back surface parameters since the electric field components are normal to the surface target. Here the charge imbalance is maintained by heating a fraction of the plasma electrons to a very large temperature. This large electron thermal pressure drives an expansion of these hot electrons, setting up a large-amplitude electrostatic field when they cross the target–vacuum interface. The accelerated ions detected behind thick targets [12] and the high-energy plasma plume emitted from the laser-irradiated surface [13] come from these "plasma sheath acceleration."

6.3 Results in Electron Beam Produced by Nonlinear Plasma Waves

The very first experiment on the FLW regime was performed on the "salle jaune" laser at Laboratoire d'Optique Appliquée (LOA), operating at 10 Hz and a wavelength of 820 nm in the CPA mode. It delivered on target energies of 1 J in 30 fs full width at half maximum (FWHM) linearly polarized pulses, whose contrast ratio was better than 10^{-6} [14]. Using a f/18 off-axis parabolic mirror, the laser beam was focused onto the sharp edge of a 3-mm supersonic helium gas jet. Since the focal spot had a waist of 18 μm, this resulted in peak intensities of up to 3×10^{18} W/cm^2.

The characterization of the electron beam was performed using an electron spectrometer, integrating current transformer (ICT), radiochromic film, and nuclear activation techniques. Typical electron beam spectra obtained at around 2.5×10^{19} cm^{-3} present a distribution with maxwellian shape (for electrons with energies below 120 MeV) with a plateau for more energetic electrons. The total charge of the electron beam was measured to be about

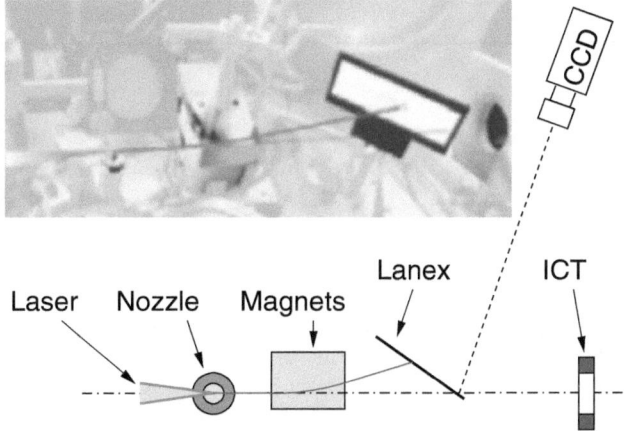

Fig. 6.2. Experimental setup with the compact device for single shot electron distribution. *Top*: picture from the experiment, *bottom*: schematic. The laser beam is focused onto a 3-mm supersonic gas jet and produces a very collimated electron beam whose spectra is measured using the compact magnet with a LANEX scintillator screen and an ICT [8]

5 nC, determined with a 10-cm-diameter ICT, installed 20 cm behind the gas jet nozzle. Subsequently, the electron beam was collimated by a 1-cm internal diameter opening in a 4-cm-thick stainless steel piece at the entrance of an electron spectrometer, which gave a collection aperture of f/100. The electron spectrum was measured with five biased silicon surfaced barrier detectors (SBD) placed in the focal plane of the electron spectrometer. By changing the magnetic field in the spectrometer from 0 to 1.5 T. it is possible to measure electrons with energies from 0 to 217 MeV. The angular distribution was measured inserting on the electron beam path a sandwich of radiochromic film and copper foil and by accumulating shots to get measurable signal on the film. By decreasing the electron density to lower values, we observed a strong saturation of the signal from the diodes measuring high-energy electrons. This was an indication that the charge at high energy had tremendously increased and that this spectrometer was no longer adapted for measuring electrons produced in this new regime. Therefore, we changed the design of our spectrometer to obtain a complete electron spectrum on a single shot basis and to lower the saturation value.

For energy distribution measurements, a 0.45-T, 5-cm-long permanent magnet was inserted between the gas jet and the LANEX screen. The LANEX screen, placed 25 cm after the gas jet, was protected by a 100-μm-thick aluminum foil in order to avoid direct exposure to the laser light. As electrons passed through the screen, energy was deposited and reemitted into visible photons, which were then imaged onto a 16 bit charged coupled device (CCD) camera. The resolution is respectively 32 and 12 MeV for 170 and 100 MeV

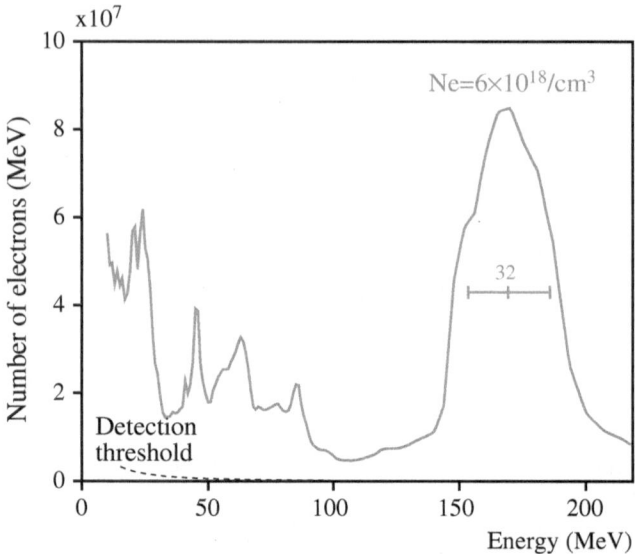

Fig. 6.3. Corresponding electron spectrum obtained at $6 \times 10^{18}\,\mathrm{cm}^{-3}$. The *dashed line* represents an estimation of the background level. The red horizontal error bars indicate the resolution of the spectrometer

energies. The charge of the electron beam was measured using an integrating current transformer placed 30 cm behind the LANEX screen. It allowed us to measure the total charge of the beam when no magnetic field was applied, and the charge above 100 MeV when the magnetic field was applied. This experimental improvement has permited to observe, in a very narrow electron density range centered at $6 \times 10^{18}\,\mathrm{cm}^{-3}$, a highly charged, 500 pC monoenergetic component at 170 MeV in the electron energy distribution $(170 \pm 20\,\mathrm{MeV})$ as it was predicted by numerical simulations.

6.4 Proton Beam Generation with Solid Targets

As already mentioned above, the same laser can be used to generate proton beams when shooting it onto overdense plasmas, for example, using solid targets. Here, the laser with an on target energy of up to 840 mJ and an FWHM duration of 40 fs was focused using a f/3 off-axis parabolic mirror. Since the focal waist was 4 μm, this resulted in peak intensities of up to $6 \times 10^{19}\,\mathrm{W/cm}^2$. For these pulses, the laser contrast ratio was, again, found to be of the order of 10^{-6}. The target, a metallic aluminum foil of 6-μm thickness, was irradiated by the laser at normal incidence.

The energy, yield, and the opening cone of generated protons were determined with CR-39 nuclear track detectors, which were partially covered with aluminum foils of varying thicknesses, which served as energy filters.

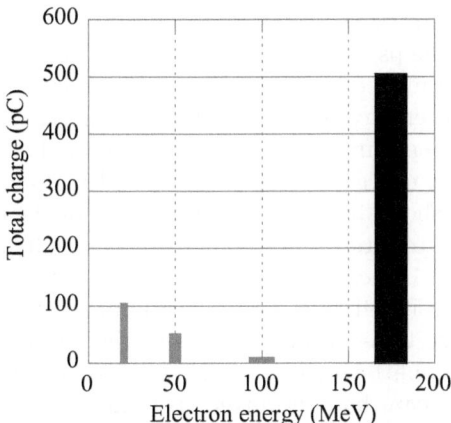

Fig. 6.4. Charge in a 10% energy bandwidth obtained at 2×10^{19} (in *gray*) and $6 \times 10^{19} \, \mathrm{cm}^{-3}$ (in *black*). Note the three orders of magnitude increased for electron energy around 175 MeV

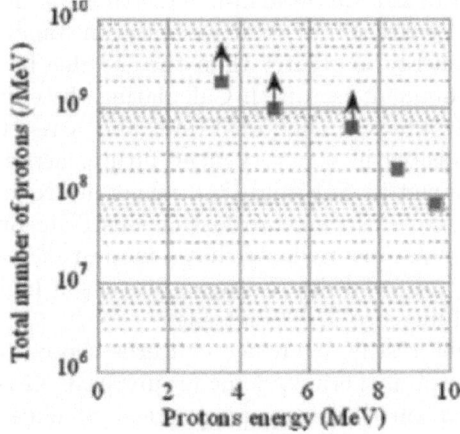

Fig. 6.5. Proton energy spectra at a laser irradiance of $6 \times 10^{19} \, \mathrm{W/cm^2}$ for a 6-μm aluminum target. The *arrow* indicates the minimum number of protons, which results in the saturation of the detectors

Figure 6.5 shows the measured proton energy distribution. Clearly, the energy of this beam reaches 10 MeV.

6.5 Perspectives

Electron beams produced by laser do not have the same properties as beams produced in conventional accelerators have. As such, they offer some complementary applications of great interest in several domains. For example, standard accelerators typically provide energetic electron bunches with a bunch duration in the ps range and an energy resolution of less than 10^{-3}. To achieve these performances, such devices are precisely designed and, hence, for a fixed electron energy only. Even though this high-energy resolution is not met in the laser–plasma accelerators approach, this approach will permit to generate tunable, energetic, high-charge, and high-quality electron beams with an extremely short duration. The shortness of the electron bunches has recently permitted to obtain interesting results on ultrafast radiation chemistry. In this pump-probe experiment, the sub-ps electron bunch was propagating through a suprasil cell containing pure liquid water producing radiolytic events. For the first time, it was probed in the sub-ps regime at LOA [15] by using a laser probe beam free of jitter. Taking benefit of the high spatial quality of the beam has permitted to radiograph a dense matter object with spatial resolution of less than 400 microns [16]. This was achieved by generating a point-like γ-ray source by bremsstrahlung radition.

Proton accelerators also produce beams with properties different from the ones produced with lasers. Even though today, the energy spectrum of this proton beam has a broad Maxwellian-like distribution, it can nevertheless be interesting for the generation of positron emission tomography (PET) radioisotopes since its energy is greater than the Q-value [few MeV for (p,n) reactions of most prominent isotopes]. Calculating the expected PET isotopes activity after an irradiation time of 30 min and a repetition rate of 1 kHz, which is indeed feasible in the very near future, activities of the order of 1 GBq can be obtained for ^{11}B and ^{18}O, which are required to separate the tracer from the inactive carrier with fast chemistry techniques [17]. Interestingly, numerical simulations indicate that a modest increase in laser intensity to 8×10^{19} W/cm^2 can result in even more protons at higher energies and can lead to a sevenfold increase in ^{18}F activity.

Another very interesting challenge concerns the use of optically induced proton beams for proton therapy. Some groups have already started to investigate this approach on the basis of numerical simulations and have shown that implementing a PW laser with a pulse duration of 30 fs and a repetition rate of 10 Hz will indeed meet the requirements for this purpose [18, 19, 20] as the dose delivered with such an adjustable proton beam spectrum within the therapeutic window (in between 60 and 200 MeV) is already expected to be beyond some few Gy/min. Importantly, this approach could provide a double benefit: (i) the size and weight of the facility is reduced, allowing a possible installation inside standard radiotherapy departments and reducing significantly the costs, and (ii) as the main beam in this "accelerator" is a laser beam, while proton generation occurs only at the end, one could also expect to reduce the size and weight of the gantries and of the associated radiation shielding [21].

6.6 Conclusion

In summary, the above-mentioned approach on optically induced particle beams has some very interesting features: (i) their accelerating field gradients are by four orders of magnitude higher than those attainable with today's standard techniques, which can consequently cut down significantly the accelerating length; (ii) the required lasers are rather compact and could become cheap in the future compared to current RF-structures; (iii) no shielding for radioprotection is required up to the point where the laser creates a plasma on the target; (iv) the same laser can be used to generate electrons or protons simultaneously; (v) the particle beams generated are of very good quality, with emittance values better than the one obtained with conventional accelerators; (vi) the particle bunches are ultrashort (durations can be less than few tens of fs); and (vii) they will be tunable.

Acknowledgments

The authors greatly appreciate the support and the quality of the "salle jaune" laser, which was ensured by the entire LOA staff. Some of the experimental data on electrons have been obtained in collaboration with Imperial College and with LULI and on protons in collaboration with the University of Strathclyde. In addition, we are indebted to the fruitful discussions with E. Lefebvre of CEA, A. Pukhov of ITP, and P. Mora of CPhT on theoretical issues. We also acknowledge the support of the European Community Research Infrastructure Activity under the FP6 "Structuring the European Research Area" program (CARE, contract number RII3-CT-2003-506395).

References

1. D. Strickland, G. Mourou: Opt. Commun. **56**(3), 219 (1985)
2. T. Tajima, J.M. Dawson: Phys. Rev. Lett. **43**, 267 (1979)
3. A. Modena, Z. Najmudin, A. Dangor, C. Clayton, K. Marsh, C. Joshi, V. Malka, B. Darrow, Danson, N. N., F. Walsh: Nature **377**, 606 (1995)
4. C. Gahn, G. Tsakiris, A. Pukhov, J. Meyer-ter Vehn, G. Pretzler, P. Thirolf, D. Habs, K. Witte: Phys. Rev. Lett. **83**, 4772 (1999)
5. V. Malka, J. Faure, J. Marquès, F. Amiranoff, J. Rousseau, S. Ranc, J. Chambaret, Z. Najmudin, B. Walton, P. Mora, A. Solodov: Phys. Plasmas **8**, 2605 (2001)
6. V. Malka, S. Fritzler, E. Levebre, M. Aleonard, F. Burgy, J.P. Chambaret, J.F. Chemin, K. Krushelnik, G. Malka, S. Mangles, Z. Najmudin, M. Pittman, J. Rousseau, J. Scheurer, B. Walton, A. Dangor: Science **298**, 1596 (2002)
7. S. Fritzler, E. Lefebvre, V. Malka, F. Burgy, A. Dangor, K. Krushelnick, S. Mangles, Z. Najmudin, J.P. Rousseau, B. Walton: Phys. Rev. Lett. **92**(16), 165006 (2004)

8. J. Faure, Y. Glinec, A. Pukhov, S. Kiselev, S. Gordienko, E. Lefebvre, J.P. Rousseau, F. Burgy, V. Malka: Nature **431**, 541 (2004)
9. Z. Najmudin, K. Krushelnick, E.L. Clark, S.P.D. Mangles, B. Walton, A.E. Dangor, S. Fritzler, V. Malka, E. Lefebvre, D. Gordon, F.S. Tsung, C. Joshi: Phys. of Plasmas **10**(5), 2071 (2003). URL: http://link.aip.org/link/?PHP/10/2071/1
10. A. Pukhov, J. Meyer-ter Vehn: Appl. Phys. B **74**, 355 (2002)
11. S.C. Wilks, W.L. Kruer, M. Tabak, A.B. Langdon: Phys. Rev. Lett. **69**, 1383 (1992)
12. R.A. Snavely, M.H. Key, S.P. Hatchett, T.E. Cowan, M. Roth, T.W. Phillips, M.A. Stoyer, E.A. Henry, T.C. Sangster, M.S. Singh, S.C. Wilks, A. MacKinnon, A. Offenberger, D.M. Pennington, K. Yasuike, A.B. Langdon, B.F. Lasnski, J. Johnson, M.D. Perry, E.M. Campbell: Phys. Rev. Lett. **85**(14), 2945 (2000)
13. E. Clark, K. Krushelnik, J. Davies, M. Zepf, M. Tatarakis, F. Beg, A. Machacek, P. Norreys, M. Santala, I. Watts, A. Dangor: Phys. Rev. Lett. **85**, 1654 (2000)
14. M. Pittman, S. Ferré, J. Rousseau, L. Notebaert, J. Chambaret, G. Chériaux: Appl. Phys. B Lasers Opt. **74**(6), 529 (2002)
15. B. Brozek-Pluska, D. Gliger, A. Hallou, V. Malka, A. Gauduel: Radiat. Phys. Chem. **72**, 149 (2005)
16. Y. Glinec, J. Faure, L. LeDain, S. Darbon, T. Hosokai, J. Santos, E. Lefebvre, J. Rousseau, F. Burgy, B. Mercier, V. Malka: Phys. Rev. Lett. **95**, 025003 (2005)
17. S. Fritzler, V. Malka, G. Grillon, J. Rousseau, F. Burgy, E. Lefebvre, E. d'Humieres, P. McKenna, K. Ledingham: Appl. Phys. Lett. **83**(15), 3039 (2003)
18. S. Bulanov, V. Khoroshkov: Plasma Phys. Rep. **28**(5), 453 (2002)
19. E. Fourkal, B. Shahine, M. Ding, J. Li, T. Tajima, C. Ma, Med. Phys. **29**(12), 2788 (2002)
20. E. Fourkal, J. Li, M. Ding, T. Tajima, C. Ma, Med. Phys. **30**(7), 1660 (2003)
21. V. Malka, S. Fritzler, E. Lefebvre, E. d'Humieres, R. Ferrand, G. Grillon, C. Albaret, S. Meyroneinc, J.P. Chambaret, A. Antonetti, D. Hulins: Med. Phys. **31**(6), 1587 (2004). URL: http://link.aip.org/link/?MPH/31/1587/1

Laser-Driven Ion Acceleration and Nuclear Activation

P. McKenna, K.W.D. Ledingham[#], and L. Robson[#]

SUPA, Department of Physics, University of Strathclyde, Glasgow, G4 0NG, UK;
[#]Also at AWE plc, Aldermaston, Reading, RG7 4PR, UK
p.mckenna@phys.strath.ac.uk

7.1 Introduction

Ion acceleration driven by intense laser–plasma interactions has been investigated since the 1970s. In early experiments with long-pulse (nanosecond) CO_2 lasers, at intensities of the order of $10^{16}\,W/cm^2$, protons were typically accelerated to tens of keV energies. They were produced with poor beam characteristics, including high transverse temperatures. The source of the protons was found to be hydrocarbon or water contamination layers on the surfaces of the laser-irradiated targets. A review of this work is provided by Gitomer et al. [1]. The introduction of chirped pulse amplification (CPA) in the late 1980s made it possible to produce high-intensity laser pulses with picosecond duration. The relativistic threshold for laser–plasma interactions was crossed at $10^{18}\,W/cm^2$, leading to collective effects in the plasma, and a renewed interest in ion acceleration. Recently, proton acceleration was observed by Clark et al. [2] and Snavely et al. [3] in short-pulse laser–plasma interactions. Protons with energies greater than 50 MeV have been measured in low divergent beams of excellent quality. This novel source of laser-driven multi-MeV energy ions has also been used to induce nuclear reactions. Rapid progress in the development of this potentially compact ion source offers intriguing possibilities for applications in isotope production for medical imaging [4], ion radiotherapy [5], ion-based fast ignitor schemes for inertial fusion energy [6], and as injectors for the next generation of ion accelerators [7].

In this chapter, some of the latest highlights on ion acceleration and ion-induced nuclear activation driven by high-intensity laser radiation are discussed. Greater emphasis is given to experimental work and, in particular, the group's research using the petawatt arm of the Vulcan laser at the Rutherford Appleton Laboratory, U.K. Two aspects of this work are reviewed. First, it is shown that ion-induced nuclear activation can be used to diagnose laser-based ion acceleration, and second, the application of laser-generated ions to induce nuclear reactions of interest to traditional areas of nuclear and accelerator physics is discussed.

P. McKenna et al.: *Laser-Driven Ion Acceleration and Nuclear Activation*, Lect. Notes Phys.
694, 91–107 (2006)
www.springerlink.com   © Springer-Verlag Berlin Heidelberg and European Communities 2006

This chapter is organized as follows: in Sect. 7.2, a brief overview of the basic physical concepts of ion acceleration in high-intensity laser–plasma interactions is presented; typical experimental arrangements are discussed in Sect. 7.3; recent experimental results are reviewed in Sect. 7.4; applications to nuclear and accelerator physics are discussed in Sect. 7.5; and future prospects are discussed in Sect. 7.6.

7.2 Basic Physical Concepts in Laser–Plasma Ion Acceleration

Ion acceleration has been demonstrated in laser–plasma experiments employing a variety of target types, including underdense (gas) targets, water droplets, cluster targets, and overdense targets, including thin foils and thicker solid targets [8]. The highest quality ion beams are produced with foil targets.

In an overdense plasma the plasma frequency (collective electron motion against the plasma ion background) is strong enough to cancel electromagnetic wave propagation. The laser–target interaction begins with the laser interacting with a preformed plasma of subcritical density (underdense). Laser light propagates only until a critical plasma density is reached (approximately 10^{21} cm^{-3}). The main laser–plasma interaction occurs at the critical density surface. The focussed laser pulse has a ponderomotive potential upon which the plasma electrons react. This has the effect of pushing electrons from regions of high laser intensity to regions of lower laser intensity, and thus electrons are accelerated because of the laser light. For peak laser intensities more than 10^{18} W/cm^2, the $v \times B$ component of the Lorentz force acting on the electrons becomes important and leads to the acceleration of electrons into the target in the laser propagation direction. The threshold for relativistic laser–plasma interactions is given by the dimensionless quantity a_0 ($= eE/\omega m_e c$, where e is the electron charge and E and ω are the electric field amplitude and frequency of the laser light, respectively). For values of $a_0 > 1$, there is a relativistic change in the electron mass, leading to collective effects in the plasma. The laser ponderomotive potential leads to a quasi-Maxwellian electron energy distribution, with temperatures (kT) in the MeV range.

The acceleration of ions, by contrast, is not achieved directly by the laser ponderomotive pressure, but is mediated by plasma processes. Electron acceleration leads to a separation of the electrons from the plasma ions, creating electrostatic fields which accelerate the ions. In this way, laser heating of the plasma leads to ion acceleration. Experimental results and numerical simulations show evidence of at least two main acceleration schemes arising from electrostatic field formation. The main acceleration schemes are summarized in Fig. 7.1. In one case, an electrostatic field is formed on the laser-irradiated surface, leading to ion acceleration from the front side of the target, and resulting in fast ions being dragged through the target, forming a beam in the forward direction [2]. In another case, referred to as the target normal sheath

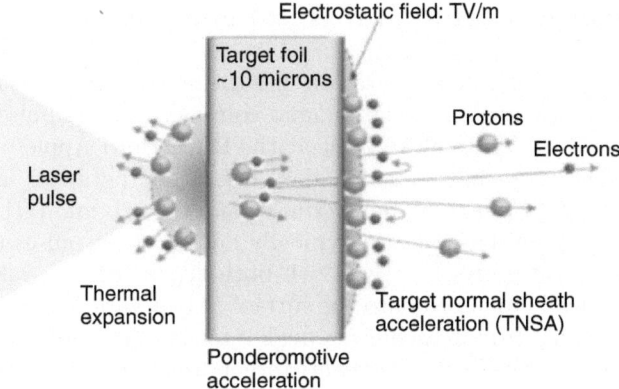

Fig. 7.1. Schematic view of the main ion acceleration processes during high-intensity laser irradiation of a thin foil target. The laser pulse, focussed from the left, creates and heats plasma on the surface of the foil. Electrons are ponderomotively driven into the foil, in the forward direction, establishing electrostatic fields on the front and rear surfaces of the foil, which lead to ion acceleration

acceleration (TNSA) model [9], the population of electrons accelerated into the target extends past the rear, nonirradiated surface, forming an electrostatic sheath on the surface, resulting in field ionization and ion acceleration. The sheath electrostatic field ($kT_{hot}/e\lambda_D$, where λ_D = Debye length) can reach values greater than 10^{12} V/m. The ions are accelerated normal to the surface, and because of the presence of a copropagating hot electron population, the ion beam is space charge neutralized. In addition to these two main mechanisms, plasma expansion at the front surface also leads to ion acceleration in the backward direction [10]. The ion beam properties strongly depend on the parameters of the charge separation. On the irradiated surface of the target, the leading edge of the laser pulse (or the amplified spontaneous emission [ASE] pedestal of background laser light) ionizes the target (at intensities of approximately 10^{12} W/cm^2) leading to an expanding plasma. Unless the target is very thin, or the ASE pedestal level is high for a sufficiently long time prior to the arrival of the main laser pulse, no preformed plasma exists on the target rear surface. The acceleration field scales inversely with the plasma scale length and hence the fields at the front of the target are lower, leading to lower energy ions.

The models outlined above have been developed in efforts to describe ion acceleration in high-temperature laser–plasma interactions. Whereas they work well for protons, the additional charge states produced for heavier ions further complicate the ionization and acceleration field dynamics. Investigations with elaborate Particle-In-Cell (PIC) codes, run on powerful parallel computers, are being employed to provide valuable modelling of this highly dynamic and complex plasma system.

7.3 Typical Experimental Arrangement

Laser–plasma–based ion acceleration has been recently investigated using large, high-energy, single-shot lasers, and compact, shorter pulse, high repetition rate lasers [8]. The Vulcan laser at the Rutherford Appleton Laboratory is typical of a large single-shot Nd:glass laser. The petawatt arm of Vulcan delivers up to 500 J pulses of 0.5-ps duration, every 20 min [11]. By contrast, compact Ti:sapphire-based lasers typically deliver short pulses (tens of fs) at high repetition rates of a few Hz. With both types of laser system, the laser pulses are typically focussed onto the surface of thin target foils to a focal spot size smaller than $10\,\mu$m, to achieve a peak intensity greater than $10^{18}\,\mathrm{W/cm^2}$.

An overview of the Vulcan petawatt target area at the Rutherford Appleton Laboratory, showing the laser compressor chamber and the target chamber in which the experiments are performed, is shown in Fig. 7.2a. Some of the ion acceleration and activation experiments performed using this laser, at intensities up to $5 \cdot 10^{20}\,\mathrm{W/cm^2}$, are discussed briefly in this chapter. Target foils, of approximately 10-μm thickness, were mounted on a heated target assembly, to facilitate resistive heating of the target to temperatures in excess of 1,000°C. The targets were irradiated at a 45° incidence angle, as shown in Fig. 7.2b. The ion diagnostics are described below.

7.3.1 Ion Diagnostics

Passive diagnostics are typically employed to diagnose ion acceleration parameters in experiments of this type. Thomson Parabola ion spectrometers, CR-39 nuclear track detectors (sensitive to ions and neutrons), dosimetry film

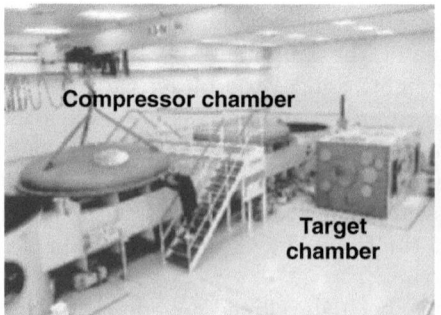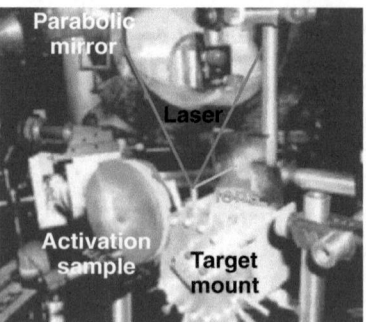

Fig. 7.2. (a) The petawatt target area of the Vulcan laser at the Rutherford Appleton Laboratory, U.K., (b) Experiment arrangement inside the petawatt target chamber for investigations into ion acceleration and nuclear activation. Laser pulses, with energy up to 500 J, are focussed using a 1.8-m focal length off-axis parabolic mirror (shown) at an angle of 45° onto the surface of thin target foils, mounted on a target wheel (shown). Accelerated ions induce nuclear activation of samples surrounding the target

(sensitive to photons, electrons, and ions), neutron time-of-flight diagnostics, and nuclear activation techniques are frequently used. Stacked dosimetry film and CR-39 are typically used to provide measurements of the spatial profile of the ion beam at different ion energies. Thomson parabola ion spectrometers have been successfully employed to measure the energies of fast protons [2, 12] and recently heavy ions [12, 13, 14], with respect to their charge-to-mass ratio. This technique involves sampling a small solid angle (typically 10^{-7} sr) of the accelerated ion beam.

Nuclear activation techniques are used to make measurements of the energy distribution of accelerated ions over a large solid angle (typically 1 sr). Measurement of the energy distribution of the full ion beam is important as the ion beam spatial profile can be influenced by self-generated electric and magnetic fields in the plasma [2]. Nuclear activation samples are positioned around the laser-irradiated target, typically along the target normal direction at both the front and the rear of the target, as illustrated in Fig. 7.3.

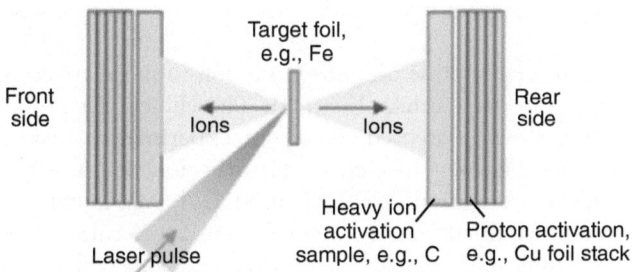

Fig. 7.3. Schematic illustrating the use of activation samples surrounding the laser-irradiated target foil to diagnose ion acceleration

Proton acceleration is usually diagnosed via proton-induced reactions in copper. Stacked copper foils (50 mm × 50 mm, with thicknesses ranging from 100 μm to 1 mm) are positioned as shown in Fig. 7.3. After irradiation, the activity of the positron emitter ^{63}Zn, produced by (p,n) reactions on ^{63}Cu, is quantified for each foil, using NaI detectors operated in coincidence and set to detect the 511 keV signature photons of positron annihilation [10]. Techniques have also been developed to enable the proton spectrum to be deduced by observing a number of reactions in a single thin Cu foil [15]. The advantage with this minimum invasive diagnostic is that it enables the fast proton beam to be used for other purposes. Other techniques under development include measurement of (p, xn) reactions in high-Z targets. These reactions offer the advantage that the cross sections are relatively high for $x = 1$ to approximately 6 and peaked at different proton energies in the range of interest for diagnosing protons produced with high-intensity lasers. As an example, cross sections for ^{206}Pb(p, xn), $x = 2$ to 5, reactions are shown in Fig. 7.4 [16].

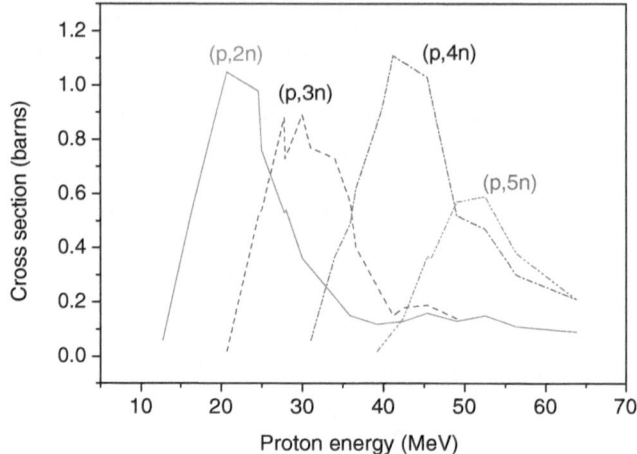

Fig. 7.4. Experimental cross sections for (p, xn), $x = 2$ to 5, reactions on ^{206}Pb [16]

Heavier ion acceleration is diagnosed via ion-induced reactions in the first activation sample in each stack (Fig. 7.3). The choice of sample material depends on the ions to be detected. In recent experiments, fusion–evaporation reactions have been chosen to characterize the ion beam, although in principle any heavy ion reaction could be used [17]. McKenna et al. [18] have used carbon activation samples to diagnose ion acceleration from Fe targets, via ^{56}Fe+^{12}C compound nucleus formation and subsequent evaporation of protons, neutrons, and α-particles. After laser irradiation, the activated sample is analyzed using a calibrated germanium detector. The residual nuclides produced in the sample are identified by the measured γ energies, intensities, and half-lifes, and quantified by correcting for detection efficiencies, gamma emission probabilities, and half-lifes.

Determination of the proton and heavy ion energy spectra involves convoluting the measured number of reactions induced, the reaction cross sections and the loss of energy as the ions propagate into the activation targets. The number of reactions is given by

$$N = D \int_{E_{\text{Thres}}}^{\infty} \sigma(E)I(E)l(E) \ \mathrm{d}E, \tag{7.1}$$

where $l(E)$ is the range of the ions at energy E, in a target of atomic density D, E_{Thres} is the threshold energy for the reaction, and $I(E)$ is the ion energy spectrum to be determined. Reaction cross sections, $\sigma(E)$, for proton-induced reactions of interest are well known. Cross sections for heavier-ion-induced fusion–evaporation reactions are typically calculated using Monte Carlo codes such as PACE-2 (projection angular-momentum coupled evaporation) [19]. McKenna et al. [17, 18] provide a fuller description of the use of the PACE-2 code and the applications of the calculated cross sections to diagnose heavy ion energy spectra.

7.4 Recent Experimental Results

A number of experimental and theoretical studies have been performed to investigate the properties of laser-accelerated ions, and in particular the acceleration of protons. As discussed in the Introduction, protons are sourced from the hydrocarbon and water contaminant layers on the target surfaces, in experiments that are largely performed in vacuum chambers at pressures of the order of 10^{-5} mbar. Heavier ions resulting from the ionization of constituent atoms in the target foils are also accelerated, but less efficiently. This is because protons, due to their high charge-to-mass ratio, outrun all other ion species and effectively screen the electrostatic acceleration fields.

7.4.1 Proton Acceleration

The energy distribution of accelerated protons is an important parameter for potential applications of this novel ion source. Figure 7.5 shows typical proton spectra, measured by proton activation of stacked Cu foils, accelerated from a 10-µm thick Al foil, irradiated by a 400 J pulse produced by the petawatt arm of the Vulcan laser. The focussed laser intensity was $2 \cdot 10^{20}$ W/cm^2.

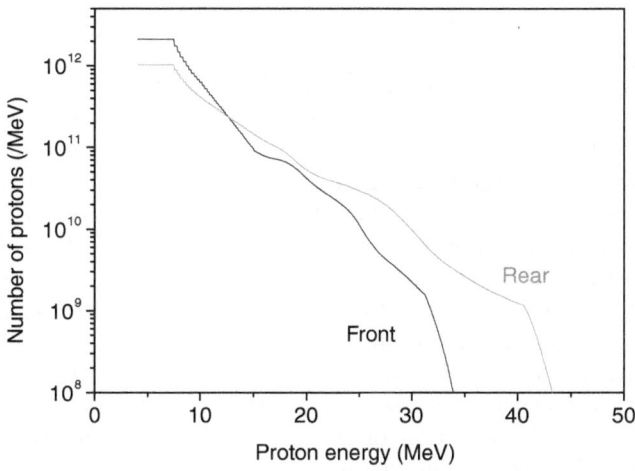

Fig. 7.5. Proton spectrum measured at the front and the rear of an Al target foil irradiated at an intensity of $2 \cdot 10^{20}$ W/cm^2. Higher energy protons are measured at the rear of the target foil (in the forward direction)

The spectra measured at both the front and the rear of the target foil have approximately exponential energy distributions with average proton energies of about 3–5 MeV. The highest energy protons are observed at the rear of the target (accelerated in the forward direction), with a sharp cutoff energy of about 45 MeV. The maximum proton energy depends on the laser irradiance ($I\lambda^2$, where I is the intensity and λ is the wavelength of the laser, respectively). This dependence has been shown to be $(I\lambda^2)^{0.4}$ up to 10^{18} W/cm^2 μm^2 [2]. For higher intensities the maximum proton energy scales as $(I\lambda^2)^{0.5}$. The proton energy distribution has also been shown to be sensitive to target parameters, such as thickness. Mackinnon et al. [20] have demonstrated that electrons accelerated during the laser interaction with the target foil recirculate in the foil, and that the mean and maximum proton energies decrease when the target thickness exceeds a thickness corresponding to the recirculation time. With thin targets the acceleration sheath at the rear of the target forms earlier, resulting in the formation of the acceleration field for a longer acceleration time.

Measurements of the spatial distribution of accelerated protons have shown that the protons are emitted in collimated beams from metallic foils whereas less well defined beams are produced with insulating targets, such as mylar [21]. The angular divergence of the collimated beams decreases with increasing proton energy. This has been shown using spatially resolving diagnostics, including CR-39 and dosimetry film [3]. The spatial distribution of accelerated protons has also been measured by contact autoradiography of activated Cu foils using Imaging Plates [22]. The plates were exposed to the activated Cu foils for about 1 hour and then scanned to yield beam-energy-differential activity distributions. Examples of the activity distributions measured on Cu foils positioned at the front and the back of the target are shown in Fig. 7.6, and are representative of the proton beam profile at increasing proton energies. It has also been shown that because the direction of the laser-accelerated ions is normal to the target rear surface, structures produced on the surface can influence the spatial distribution of the accelerated ion beam [23]. It follows that by shaping the target rear surface in the form of a hemisphere the proton beam can be focussed. This has been demonstrated experimentally [23] and theoretically [9]. Furthermore, experiments at the Lund Laser Centre have shown that under certain conditions the emitted proton beam deviates from target normal direction, toward the laser forward direction [24]. The angle of deviation was found to change with the level and timing of the ASE pedestal with respect to the main laser pulse, and importantly the angle of deviation was shown to increase with proton energy. This result points to a controllable spatial separation of the proton beam in energy at the source and could have important implications for the many potential applications of laser-based proton sources [24].

An important property of this novel source of protons is the beam quality. As ion acceleration at the rear of the target takes place from a cold, initially unperturbed surface, a low beam emittance results. By measuring the spatial

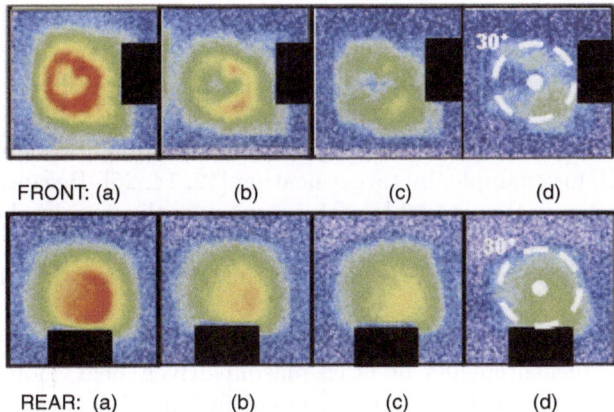

FRONT: (a)   (b)   (c)   (d)

REAR: (a)   (b)   (c)   (d)

Fig. 7.6. Spatial distribution of proton-induced activity in Cu foils stacked at the front and rear sides of a 10-μm-thick Al target irradiated by a 170 J Vulcan laser pulse. Imaging plates were exposed to the activated Cu foils for about 1 hour. The level of activity is color coded in the sequence: red (highest); orange; yellow; green; blue (lowest). The measured activity is from proton-induced reactions [principally $^{63}Cu(p,n)^{63}Zn$] and therefore represents the proton beam spatial profile in differing energy ranges stopped in each foil, as follows: (**a**) 4.0 (reaction threshold) to 5.8 MeV; (**b**) 5.8–9.0 MeV; (**c**) 9.0–14.5 MeV; and (**d**) 14.5–18.5 MeV. The markings shown in (d) are valid for (a) to (d), where the white dot corresponds to the target normal direction, and the ring corresponds to a beam with a 30° opening cone angle. The spatial profile of the beam measured at the rear is quite uniform in comparison to the ring-like distribution measured at the front. The ring structure is thought to result from the deflection of protons in magnetic fields of the order of 10^7 Gauss, established within the plasma

distribution of protons accelerated from a structured target surface, Cowan et al. [25] have shown experimentally that, for protons of up to 10 MeV, the transverse emittance is as low as 0.004 mm mrad. This corresponds to a 100-fold better quality, more laminar beam, than produced from typical RF accelerators. A further important parameter of laser-accelerated protons is that they are produced in short bunches. As ions are accelerated for as long as the space charge separation in the plasma driven by the laser is maintained, the ion beam pulse duration at the source will be of the order of the laser pulse duration.

7.4.2 Heavier Ion Acceleration

In addition to multi-MeV proton acceleration, heavier ion acceleration has also been investigated [7, 12, 13, 17]. Carbon, aluminum, fluorine, and lead ions have been observed with energies up to approximately 5 MeV/nucleon from thin foils irradiated with laser intensities of 10^{19} to 10^{20} W/cm². Hegelich et al. [12] have indicated that field ionization is the dominant ionization mechanism and that recombination and collisional ionization have minor contributions. Recent experimentation has also shown that heavy ions are more efficiently accelerated when hydrogen-containing surface contaminants are removed from the target foil, for example, by target heating [12, 14, 26]. Because of the large number of charge states available with heavy ions, the measured energy spectra of different ion species has been shown to provide additional information regarding the spatiotemporal evolution of the accelerating field, not available in the proton signal [12].

Nuclear activation techniques have only recently been developed and employed for measurements of laser–plasma–driven heavy ion acceleration [17]. This was achieved by measurements of fusion–evaporation reactions, as discussed above. The technique not only facilitates spatially integrated measurements of heavy ion energies, but by contact autoradiography could potentially also be used to make spatially resolved measurements.

The nuclear activation techniques described have been applied to compare proton and heavier ion acceleration, from the same laser shots, with heated and unheated targets [18], using the arrangement shown in Fig. 7.3. Activation measurements in stacked Cu foils were used to diagnose proton acceleration, and acceleration of Fe ions was diagnosed via measurements of [Fe+C] fusion–evaporation reactions in C samples. Effective removal of the hydrogen contaminants was achieved by resistively heating the target foils to temperatures in excess of 850°C. Part of a measured γ emission spectrum from the carbon sample positioned at the front side of the target is shown in Fig. 7.7. The laser intensity for these shots was approximately 3×10^{20} W/cm².

The ion energy spectra deduced by convoluting the number of each reaction observed, the stopping ranges in the sample and the calculated cross sections, as discussed above, are shown in Fig. 7.8. With the Fe target foil unheated, Fe ions were accelerated up to approximately 450 MeV. The conversion efficiency from laser energy to Fe ion acceleration (with energy above approximately 150 MeV) was determined to be approximately 0.8%. When the target was heated, the numbers of Fe ions accelerated over the observed energy range was up to an order of magnitude higher. Fe ions are accelerated to greater than 600 MeV, and with energy conversion efficiency of approximately 4.2%. These spatially integrated ion flux measurements clearly illustrates that removal of contaminants by target heating increases the efficiency of heavier ion acceleration considerably, in line with results using ion spectrometers to sample small solid angles [12, 14].

The acceleration of protons from both the front and the rear surfaces of the heated and unheated Fe target foils was also diagnosed for the same laser shots (Fig. 7.8). Proton acceleration was suppressed by resistive heating of the targets. With the unheated target, up to 10^{12} protons per shot with a maximum energy >40 MeV were observed. The energy conversion efficiency

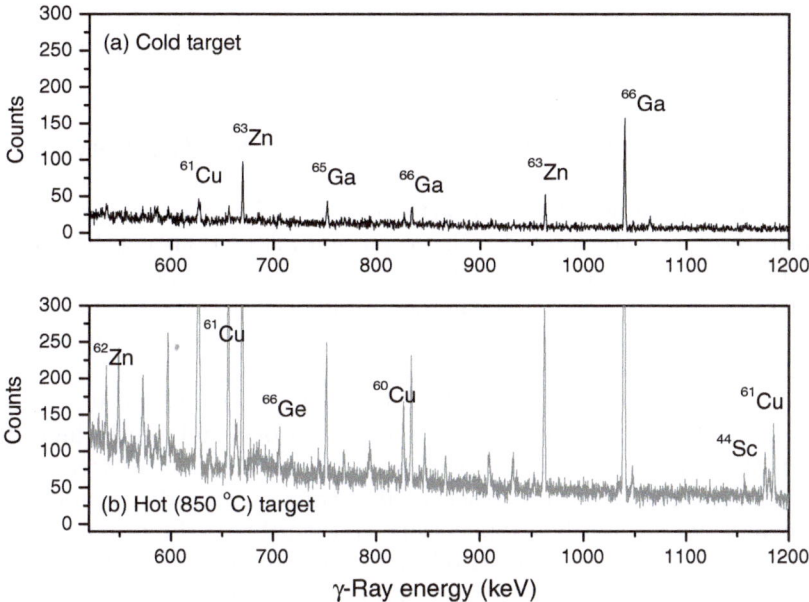

Fig. 7.7. Signature γ-ray peaks resulting from ion-induced reactions in carbon activation samples positioned at the front of a 100-μm-thick Fe foil target which was (**a**) cold and (**b**) heated to 850°C for 30 min prior to the laser shot to remove hydrogen-containing contaminants

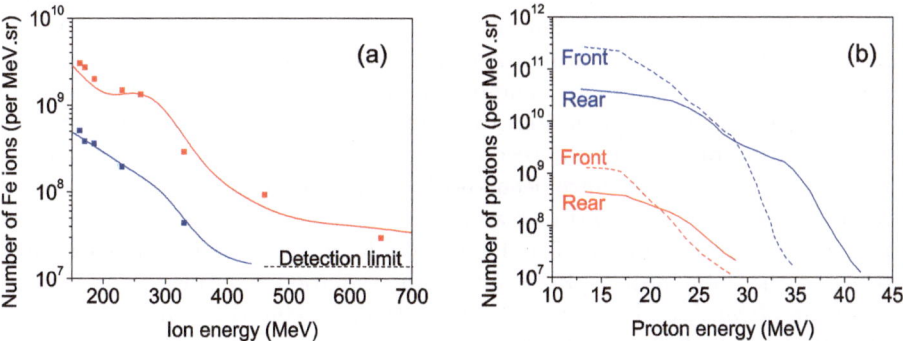

Fig. 7.8. (**a**) Energy spectrum of Fe ions deduced by ion activation of C, from unheated (blue) and heated (red) Fe foil target. (**b**) Proton energy spectra measured by proton activation of Cu, from the same unheated (blue) and heated (red) target. Target heating is observed to reduce the flux of protons and lead to more efficient acceleration of heavier ions

to protons was measured to be approximately 7%. Target heating resulted in a reduction of the numbers of fast protons to between 10^9 and 10^{10} per shot.

7.5 Applications to Nuclear and Accelerator Physics

In addition to the use of nuclear activation techniques as a diagnostic of ion acceleration, ions accelerated in laser–plasma interactions can be applied to investigate nuclear reactions of interest to the traditional fields of nuclear and accelerator science. This novel, and potentially compact, laser-based source of multi-MeV ions could benefit the many users of conventional ion accelerator technology.

The requirements for some applications, for example, medical isotope production [4, 10] are satisfied by the production of a sufficiently high flux of ions with energy above the target reaction threshold energy, whereas many applications, such as ion radiotherapy [5], require the production of beams of monoenergetic ions. Other applications can benefit by the production of ion beams with specific energy distributions. An example of the latter, discussed below, is the application of the typically broad energy distribution of protons accelerated in a laser–plasma interaction, to the study of nuclear reactions of interest for the development of accelerator-driven systems [27].

7.5.1 Residual Isotope Production in Spallation Targets

The proposed development of the accelerator-driven systems as a source of neutrons is based on the physics of spallation reactions [28]. A proton-induced spallation reaction occurs when a proton with energy of the order of a GeV, inelastically collides with a high-Z target nucleus, and "spalls" or knocks out protons, neutrons, and pions. These high-energy secondary particles typically cause further spallation reactions by intranuclear cascade, in the first stage of the spallation process. In the second stage of spallation, the excited nucleus, from which the particles have been knocked out, deexcites by evaporating large numbers of low-energy particles and/or by fission. The two stages of spallation are illustrated in Fig. 7.9. Because a large percentage (approximately 60%) of the total proton flux emitted in spallation reactions have energies less than 50 MeV, and the secondary reactions produced by these low-energy protons have high cross sections, they contribute significantly to the residual radioactivity produced in a spallation target. As this activity defines the radioinventory and sets the handling limit on spallation targets, the study of these secondary reactions is important to the development of accelerator-driven systems. Laser-based sources of multi-MeV protons have recently been applied to investigate residual isotope production in spallation targets [29].

Cross sections for low-energy proton production via the intranuclear cascade and evaporation processes are shown in Fig. 7.10, for three incident proton energies, 0.8, 1.2, and 2.5 GeV, on a Pb spallation target [30]. From

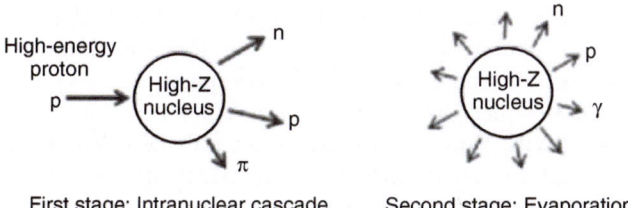

Fig. 7.9. Schematic illustration of spallation reactions. A high-energy (GeV) proton knocks out protons, neutrons, and pions from a high-Z nucleus. The remaining excited nucleus evaporates large numbers of particles including protons and neutrons with energies in the tens of MeV range. These secondary particles create further nuclear reactions, contributing to the residual activity of the spallation target

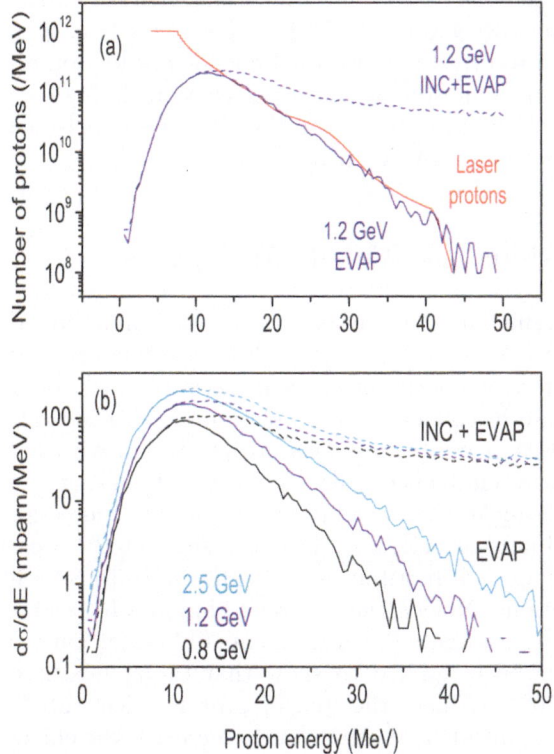

Fig. 7.10. (a) Energy spectrum (*red line*) of protons accelerated using the Vulcan petawatt laser. (b) Calculated cross sections for proton production via spallation–evaporation, EVAP (*solid lines*), and spallation-intranuclear cascade plus evaporation, INC+EVAP (*broken lines*), reactions in 0.8, 1.2, and 2.5 GeV p+Pb interactions [30]. The calculated 1.2 GeV spectra (*blue lines*) are shown normalized to the laser-generated proton spectrum in (a) – excellent agreement is observed with the EVAP spectrum for energies greater than 12 MeV

neutron multiplicity considerations, the optimum incident proton energy for spallation is approximately 1.1 GeV, and therefore the shapes of the 1.2 GeV p+Pb emitted proton energy spectra are compared to a typical energy spectrum of protons produced by interaction of Vulcan laser pulses (300 J, 0.7 ps, 3×10^{20} W/cm^2) with thin (10 μm) Al target foils (Fig. 7.10). McKenna et al. [29] observed that above 12 MeV there is excellent agreement between the shape of the measured laser–plasma proton energy distribution and the calculated energy spectrum of protons emitted in spallation–evaporation reactions. Furthermore, the lower energy threshold for the dominant reactions is approximately 12 MeV. Therefore, the characteristic energy distribution of protons accelerated with the Vulcan petawatt laser was applied to experimentally model the isotopic distribution of residual radioisotopes produced in a Pb sample – a representative spallation target [29].

This work demonstrated that the broad energy distribution feature of beams of protons accelerated in high-temperature laser–plasma interactions could be applied to investigate residual nuclide production resulting from protons produced in spallation–evaporation reactions in high-Z targets. The isotopic distribution data produced can thus be used to help benchmark nuclear codes in the low-energy (MeV) regime.

7.6 Conclusions and Future Prospects

Some recent results in the acceleration of ions and the production of ion-induced nuclear reactions, driven by high-intensity lasers, have been reviewed. The production of acceleration fields in laser–plasmas, orders of magnitude larger than those in conventional accelerators, has been established. Bright, picosecond pulses of multi-MeV ions are produced, with low transverse and longitudinal beam emittance. These unique and conventionally unattainable beam properties make this a highly interesting ion source for applications.

Ion-induced nuclear activation has been shown to be a useful diagnostic of the energy and spatial distribution of beams of protons accelerated in laser–plasma interactions. Nuclear activation techniques have also been developed to make spatially integrated measurements of heavier ion acceleration. These techniques have been applied to show that target heating to temperatures in excess of 850°C reduces the flux of protons accelerated by greater than two orders of magnitude and significantly increases the efficiency of heavy ion acceleration. Nuclear activation techniques benefit from a large dynamic range and insensitivity to electrical noise generated by the laser–plasma interaction, as counting is carried out offline.

In addition to providing new diagnostic capabilities for laser–plasma physics, nuclear activation driven by lasers facilitates the investigation of nuclear reactions without recourse to nuclear reactors or conventional accelerator technology. A beam of protons with a broad energy distribution, for example, has been applied to investigate residue production due to low-energy reactions in spallation targets. High-intensity laser-based sources of neutrons and gamma radiation have also been used to induce nuclear reactions and make measurements of cross sections relevant to transmutation physics [31, 32].

Significant advances have been made recently in laser technology. The repetition rate of terawatt-class lasers has increased substantially, the pulse duration of terawatt lasers is also decreasing, and introduction of adaptive optics, such as deformable mirrors, will ensure that laser radiation can be focussed almost to the diffraction limit (a single wavelength). As laser intensities continue to increase, ions with increased energy will be produced. At intensities of the order of 10^{24} W/cm^2 protons will be accelerated to relativistic velocities. Protons with upper energies in the range 100 MeV to 1 GeV will facilitate investigation of proton-induced fission and spallation. As laser intensities increase, the electric field in the focussed laser pulse should reach values high enough to directly affect the nucleus. This will enable intense lasers to become a valuable tool for nuclear physics.

Advances in target design and engineering may also improve ion beam quality. Esirkepov et al. [33] used 3D PIC simulations of laser irradiation of double-layer targets, consisting of a relatively thick first layer of high-Z material coated by a very thin film of low-Z atoms, to suggest that the accelerated proton energy distribution could be made quasi monoenergetic. In addition, by altering the geometry of the target foil, either by using preshaped targets or by using controlled low-temperature shock waves to modify the target conditions, it is possible to influence the directionality and properties of the accelerated ions [24].

Finally, intense lasers can be used to produce not only pulses of fast ions but also high-energy neutrons, electrons, and gamma radiation. Furthermore, these pulses of particles and radiation can be synchronized to picosecond timescales, which may enable lasers to find truly unique applications in the production and investigation of very short-lived isotopes.

Acknowledgments

The contribution of colleagues at the University of Strathclyde, the Central Laser Facility-Rutherford Appleton Laboratory, the University of Paisley, Queen's University Belfast, the University of Glasgow, the Institute of Transuranium Elements, Karlsruhe, and Imperial College London to this research is highly appreciated. PMcK gratefully acknowledges the award of a Royal Society of Edinburgh Personal Fellowship. We gratefully acknowledge D. Hilscher and C.-M. Herbach for fruitful communications regarding cross sections for spallation processes.

References

1. S. Gitomer, R. Jones, F. Begay, A. Ehler, J. Kephart, R. Kristal: Phys. Fluids **29**, 2679 (1986)
2. E. Clark, K. Krushelnick, J. Davies, M. Zepf, M. Tatarakis, F. Beg, A. Machacek, P. Norreys, M. Santala, I. Watts, A. Dangor: Phys. Rev. Lett. **84**, 670 (2000)
3. R. Snavely, S. Hatchett, T. Cowan, M. Roth, T. Phillips, M. Stoyer, E. Henry, C. Sangster, M. Singh, S. Wilks, A. Mackinnon, A. Offenberger, D. Pennington, K. Yasuike, A. Langdon, B. Lasinski, J. Johnson, M. Perry, E. Campbell: Phys. Rev. Lett. **85**, 2945 (2000)
4. K. Ledingham, P. McKenna, T. McCanny, S. Shimizu, J. Yang, L. Robson, J. Zweit, J. Gillies, J. Bailey, G. Chimon, R. Singhal, M. Wei, S. Mangles, P. Nilson, K. Krushelnick, M. Zepf, R. Clarke, P. Norreys: J. Phys. D Appl. Phys. **37**, 2341 (2004)
5. S.V. Bulanov, T. Esirkepov, V. Khoroshkov, A. Kunetsov, F. Pegoraro: Phys. Lett. A **299**, 240 (2002)
6. M. Roth, T. Cowan, M. Key, S. Hatchett, C. Brown, W. Fountain, J. Johnson, D. Pennington, R. Snavely, S. Wilks, K. Yasuike, H. Ruhl, F. Pegoraro, C. Bula, E. Campbell, M. Perry, H. Powell: Phys. Rev. Lett. **86**, 436 (2001)
7. K. Krushelnick, E. Clark, R. Allott, F. Beg, C. Danson, A. Machacek, V. Malka, Z. Najmudin, D. Neely, P. Norreys, M. Salvati, M. Santala, M. Tatarakis, I. Watts, M. Zepf, A. Dangor: IEEE Transact. Plasma Sci. **28**, 1184 (2000)
8. J.T. Mendonca, J.R. Davies, M. Eloy: Meas. Sci. Technol. **12**, 1801 (2001)
9. S.C. Wilks, A. Langdon, T. Cowan, M. Roth, M. Singh, S. Hatchett, M. Key, D. Pennington, A. Mackinnon, R. Snavely: Phys. Plasmas **8**, 542 (2001)
10. I. Spencer, K. Ledingham, R. Singhal, T. McCanny, P. McKenna, E. Clark, K. Krushelnick, M. Zepf, F. Beg, M. Tatarakis, A. Dangor, P. Norreys, R. Clarke, R. Allott, I. Ross: Nucl. Instrum. Methods Phys. Res. B **183**, 449 (2001)
11. C. Danson, P. Brummitt, R. Clarke, J. Collier, B. Fell, A. Frackiewicz, S. Hancock, S. Hawkes, C. Hernandez-Gomez, P. Holligan, M. Hutchinson, A. Kidd, W. Lester, I. Musgrave, D. Neely, D. Neville, P. Norreys, D. Pepler, C. Reason, W. Shaikh, T. Winstone, R. Wyatt, B. Wyborn: IAEA J. Nucl. Fusion, **44**, 239 (2004)
12. M. Hegelich, S. Karsch, G. Pretzler, D. Habs, K. Witte, W. Guenther, M. Allen, A. Blazevic, J. Fuchs, J. Gauthier, M. Geissel, P. Audebert, T. Cowan, M. Roth: Phys. Rev. Lett. **89**, 085002 (2002)
13. E.L. Clark, K. Krushelnick, M. Zepf, F. Beg, M. Tatarakis, A. Machacek, M. Santala, I. Watts, P. Norreys, A. Dangor: Phys. Rev. Lett. **85**, 1654 (2000)
14. M. Zepf, E. Clark, F. Beg, R. Clarke, A. Dangor, A. Gopal, K. Krushelnick, P. Norreys, M. Tatarakis, U. Wagner, M. Wei: Phys. Rev. Lett. **90**, 064801-1 (2003)
15. J. Yang, P. McKenna, K. Ledingham, T. McCanny, S. Shimizu, L. Robson, R. Clarke, D. Neely, P. Norreys, M. Wei, K. Krushelnick, P. Nilson, S. Mangles, R. Singhal: Appl. Phys. Lett. **84**, 675 (2004)
16. R.E. Bell, H.M. Skarsgard: Can. J. Phys. **34**, 745 (1956)
17. P. McKenna, K. Ledingham, T. McCanny, R. Singhal, I. Spencer, M. Santala, F. Beg, A. Dangor, K. Krushelnick, M. Takarakis, M. Wei, E. Clark, R. Clarke, K. Lancaster, P. Norreys, K. Spohr, R. Chapman, M. Zepf: Phys. Rev. Lett. **91**, 075006 (2003)

18. P. McKenna, K. Ledingham, J. Yang, L. Robson, T. McCanny, S. Shimizu, R. Clarke, D. Neely, K. Krushelnick, M. Wei, P. Norreys, K. Spohr, R. Chapman, R. Singhal: Phys. Rev. E, **70**, 036405 (2004)
19. A. Gavron: Phys. Rev. C **21**, 230 (1980)
20. A.J. Mackinnon, Y. Sentoku, P. Patel, D. Price, S. Hatchett, M. Key, C. Andersen, R. Snavely, R. Freeman: Phys. Rev. Lett. **88**, 215006 (2002)
21. J. Fuchs, T. Cowan, P. Audebert, H. Ruhl, L. Grémillet, A. Kemp, M. Allen, A. Blazevic, J.-C. Gauthier, M. Geissel, M. Hegelich, S. Karsch, P. Parks, M. Roth, Y. Sentoku, R. Stephens, E. Campbell: Phys. Rev. Lett. **91**, 255002 (2003)
22. Imaging Plates: details online at http://home.fujifilm.com/products/science/ip/
23. M. Roth, M. Allen, P. Audebert, A. Blazevic, E. Brambrink, T. Cowan, J. Fuchs, J.-C. Gauthier, M. Geissel, M. Hegelich, S. Karsch, J. Meyer-ter-Vehn, H. Ruhl, T. Schlegel, R. Stephens: Plasma Phys. Control. Fusion **44**, B99 (2002)
24. F. Lindau, O. Lundh, A. Persson, P. McKenna, K. Osvay, D. Batani, and C.-G. Wahlström: Physical Review Letters, **95**, 175002 (2005)
25. T. Cowan, J. Fuchs, H. Ruhl, A. Kemp, P. Audebert, M. Roth, R. Stephens, I. Barton, A. Blazevic, E. Brambrink, J. Cobble, J. Fernandez, J.-C. Gauthier, M. Geissel, M. Hegelich, J. Kaae, S. Karsch, G. Le Sage, S. Letzring, M. Manclossi, S. Meyroneinc, A. Newkirk, H. Pepin, N. Renard-LeGalloudec: Phys. Rev. Lett. **92**, 204801 (2004)
26. P. McKenna, K. Ledingham, T. McCanny, R. Singhal, I. Spencer, E. Clark, F. Beg, K. Krushelnick, M. Wei, R. Clarke, K. Lancaster, P. Norreys, J. Galy, J. Magill: Appl. Phys. Lett. **83**, 2763 (2003)
27. C. Rubbia, J. Rubio, S. Buono, F. Carminati, N. Fietier, J. Galvez, C. Geles, Y. Kadi, R. Klapisch, P. Mandrillon, J. Revol, C. Roche: CERN/AT/95-44(ET)
28. N. Watanabe: Rep. Prog. Phys. **66**, 339 (2003)
29. P. McKenna, K. Ledingham, S. Shimizu, J. Yang, L. Robson, T. McCanny, J. Galy, J. Magill, R. Clarke, D. Neely, P. Norreys, R. Singhal, K. Krushelnick, M. Wei: Phys. Rev. Lett. **94**, 084801 (2005)
30. D. Hilscher, C.-M. Herbach: private communication
31. K. Ledingham, J. Magill, P. McKenna, J. Yang, J. Galy, R. Schenkel, J. Rebizant, T. McCanny, S. Shimizu, L. Robson, R. Singhal, M. Wei, S. Mangles, P. Nilson, K. Krushelnick, R. Clarke, P. Norreys: J. Phys. D Appl. Phys. **36**, L79 (2003).
32. J. Magill, H. Schwoerer, F. Ewald, J. Galy, R. Schenkel, R. Sauerbrey: Appl. Phys. B Lasers Opt. **77**, 387 (2003)
33. T.Z. Esirkepov, S. Bulanov, K. Nishihara, T. Tajima, F. Pegoraro, V. Khoroshkov, K. Mima, H. Daido, Y. Kato, Y. Kitagawa, K. Nagai, S. Sakabe: Phys. Rev. Lett. **89**, 175003 (2002)

8

Pulsed Neutron Sources
with Tabletop Laser-Accelerated Protons

T. Žagar[1], J. Galy[2], and J. Magill[2]

[1]Jožef Stefan Institute, Jamova 39, 1000 Ljubljana, Slovenia
tomaz.zagar@ijs.si
[2]European Commission, Joint Research Centre, Institute for Transuranium Elements, Postfach 2340, 76125 Karlsruhe, Germany
Joseph.Magill@cec.eu.int

Abstract. Neutron production rates using laser-accelerated protons from high-energy single-shot laser (giant pulse laser) and low-energy high-repetition tabletop laser systems are compared. With the VULCAN giant pulse laser, more than 10^9 neutrons per shot were produced in a nanosecond pulse through (p,xn) reactions with lead. In contrast, a current state-of-the-art tabletop laser theoretically can produce 10^6 to 10^7 neutrons per second in repetitional nanosecond pulses. It is estimated that next-generation tabletop lasers currently under construction will be capable of producing nanosecond neutron pulses at a rate of 10^{10} neutrons per second.

8.1 Introduction

The ability to induce a variety of nuclear reactions with high-intensity lasers has been demonstrated recently in several laboratories [1, 2]. Laser-induced activation, fission, fusion, and transmutation [3, 4, 5, 6, 7, 8, 9] have been demonstrated without recourse to reactors or large-scale particle accelerators. These astonishing results came after a variety of technological breakthroughs in the field of high-intensity lasers in the last decade. CPA (Chirped Pulse Amplification [10]), being one of the examples, made it possible to obtain laser intensities above $10^{19}\,\mathrm{W/cm^2}$ in the focal point. It was observed in many laboratories that such high-intensity laser beam focussed on a surface of a thin solid target can generate collimated jets of high-energy electrons, protons, and heavy ions. The production of protons with energies above 1 MeV was measured for the first time with the VULCAN laser at the RAL (Rutherford Appleton Laboratory) [11] already in 1994. Nowadays, also physicists using high-power and high-intensity tabletop lasers from several other laboratories (LLNL – Lawrence Livermore National Laboratory [12], LULI – Laboratoire pour l'Utilisation des Lasers Intenses, Palaiseau [13], CUOS – Center for Ultrafast Optical Science, Michigan [14], LOA – Laboratoire d'Optique Appliquée, Palaiseau [15]) have characterized this proton source.

T. Žagar et al.: *Pulsed Neutron Sources with Tabletop Laser-Accelerated Protons*, Lect. Notes Phys. **694**, 109–128 (2006)
www.springerlink.com   © Springer-Verlag Berlin Heidelberg and European Communities 2006

There are many exciting applications for fast protons. Fast protons could be used for proton radiography, for generation of nearly monoenergetic sources of high-energy γ-rays, for generation of extremely short-lived isotopes, and much more. The objective of this work is to cover the production of neutrons using laser-accelerated protons and not to present the complete picture of proton acceleration using high-intensity lasers. Taking into account high proton energies and high laser-to-proton energy conversion efficiency, the (p,xn) reactions seem to be an interesting option for a novel compact neutron source [16]. Again (p,xn) reactions are not the only way to produce neutrons using high-intensity lasers. Another possibility to generate neutrons is to use laser-generated deuterium plasma source. A nearly pure monoenergetic spectrum of fusion neutrons from the D + D nuclear reaction can be obtained by irradiating deuterium atomic clusters to drive fusion between hot deuterons from neighboring cluster explosions. Cluster fusion sources are capable of yields up to 10^5 fusion neutrons per Joule of incident laser energy [17]. As will be shown here, higher neutron yields can be obtained using laser-accelerated protons to generate neutrons in solid targets.

We do not need to say that neutron sources have an extremely broad range of applications. Their applications stared in the middle of last century, and after more than 50 years of development they are quite widespread ranging from neutron activation analysis [18, 19] through nuclear geophysics [20] and applications in nuclear medicine to neutron radiography and active neutron interrogation methods used in homeland security [21, 22]. In all those fields pulsed, compact, strong, and mobile neutron source would be of great advantage. And laser-driven neutron source has great potential of being more compact and portable than accelerator-based sources [17].

8.2 Recent Proton Acceleration Experiments

Laser-driven ion acceleration is a cascade process. For very intense laser fields, the ionization of target atoms is instantaneous, transforming the target material into plasma. The high-intensity laser beam (intensities above 10^{19} W/cm^2) focussed on a surface of a thin solid target generates plasma with temperatures greater than 10 billion degrees (10^{10} K). Once the electrons become free and are accelerated by the laser field, the secondary process of ion acceleration takes place. Two types of electromagnetic forces act on the free electrons: the electrical force and the ponderomotive force. The electrical force is due to the direct action of the electrical field on the electrons and drives electrons into fast oscillations around their initial positions and consequently to electron heating. On the other hand, the radiation pressure, or ponderomotive force, pushes electrons away from the region occupied by the laser pulse [23]. For very short pulses, this force can become very important and resulting acceleration will tend to push the electrons in front of the laser pulse. And so after the passage of the laser pulse, the plasma region becomes positively charged.

After this initial charge separation, several proton acceleration mechanisms can take place depending on the laser intensity, laser pulse characteristics, and target characteristics. These acceleration mechanisms will be briefly described in next paragraphs.

The emission of fast protons from solid targets (foils) has been measured in numerous experiments since the 1960s. During this time, a broad range of laser intensities was used for ion production (from 10^{14} to 10^{20} W/cm^2 and more in the last years). Energetic protons and ions generated by focussing nanosecond pulses on solid targets at intensities of 10^{14} to 10^{16} W/cm^2 are emitted into a large solid angle. They exhibit strong trajectory crossings and a broad energy spectrum with typical ion temperatures of 100 keV per nucleon. This scenario is totally different when the ion acceleration is caused by femtosecond (fs) laser pulses. When these are focussed on thin foils targets at intensities above 10^{19} W/cm^2, proton beams are observed, which exhibit two new distinguished features: first, the temperature of these protons is in the order of few MeV per nucleon with maximum energies between 5 and 50 MeV per nucleon; second, well-collimated beams of protons, within the opening angle smaller than 20 degrees, are generated in the target normal direction on front and back sides of the target foil.

Recent experimental studies have demonstrated the production of beams of protons with energies up to 58 MeV, using a laser with wavelength of approximately 1 μm and intensity of 3×10^{20} W/cm^2 [24]. Not only the high energy but also high efficiency of laser-to-proton energy conversion is of particular interest. In the same experiment the conversion efficiency of laser into fast proton energy was approximately 12%.

Detected protons originate from water and hydrocarbon molecules adsorbed at the target foil surfaces due to the presence of water and pump oil vapor in the target area. The complete physical understanding of acceleration mechanisms for the most energetic protons is still somewhat debated; however, there is general consensus that two general acceleration scenarios are able to explain the occurrence of collimated MeV proton beams [25]. They may come from the front surface, from the rear surface, or even both mechanisms acting simultaneously. Even if the physics of the acceleration mechanisms is not fully understood, the correlation between laser intensity and maximum proton energy was deduced empirically from existing experimental data. Measured maximal proton energy scales linearly in logarithmic scale with laser intensity multiplied by wavelength squared as can be seen on Fig. 8.1. This relation can be written as $(I\lambda^2)^{-\alpha}$, where I is the laser intensity in W/cm^2, λ is the laser wavelength, and α is, according to various authors, between 0.3 and 0.6. Data points presented on this figure are a compilation of our results and analyses originally made by Mendonça [23], Clark [26], and Spencer [27]. Results of our measurements described later in this work are highlighted on this figure. Clearly, there are large shot-to-shot variations in the data, since different measurement techniques, laser contrasts, pulse durations, and foil thicknesses were used in different experiments.

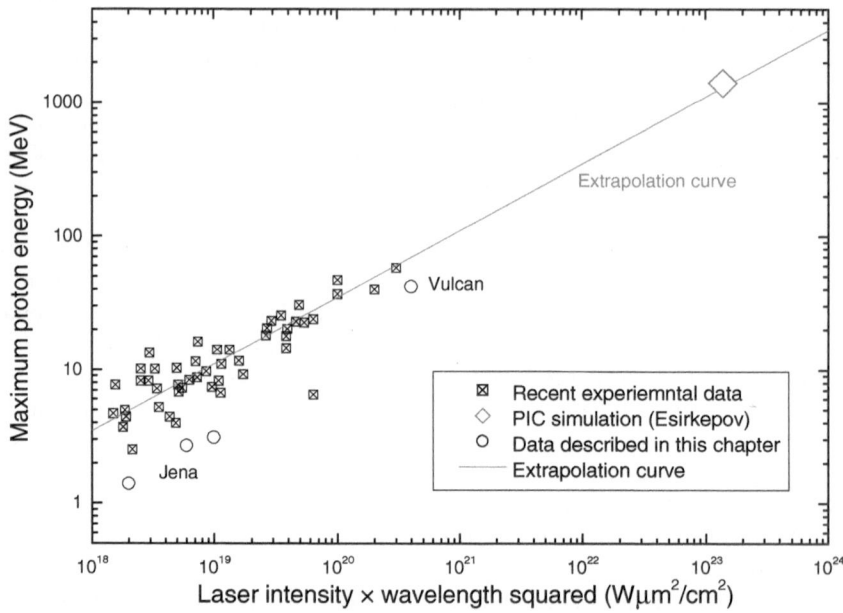

Fig. 8.1. Maximum proton energy plotted as a function of laser irradiance ($I\lambda^2$). Recently published proton acceleration results are marked with crossed squares and they lie close to extrapolation curve determined by Clark [26], Mendonça [23], and Spencer [27]. These data points include experimental results obtained on different single-shot and tabletop laser facilities around the world. PIC simulation point presented by Esirkepov [28] (marked with a diamond) lies on the extrapolation curve. Data for the experiments described in this work are marked with circles

As mentioned before, Clark, Mendonça, and Spencer came to the result that the maximum proton energy scales as

$$E_{\max} = \sqrt{I_L \lambda^2} \cdot 3.5 (\pm 0.5) \times 10^{-9} \,, \tag{8.1}$$

where $E_{\max}$ is maximum proton energy in MeV and $I_L \, \lambda^2$ is the laser intensity multiplied by laser wavelength in $W\mu m^2/cm^2$. This relation is supported by many experimental results in the laser intensity range between 10^{18} and $4 \times 10^{20}\,W\mu m^2/cm^2$. The extrapolation of this relation to higher intensities, as it is presented on Fig. 8.1, can be supported only by numerical simulations. One of such PIC simulations presented recently is the simulation of plasma and laser interaction for intensities as high as $10^{23}\,W\mu m^2/cm^2$ done by Esirkepov [28]. His simulations of the "laser piston" principle accelerated ions to energies above 1.4 GeV. As we can see, this data point falls on the extrapolation line which we use to correlate maximum proton energies and laser intensities in this work. These intensities are not in the reach of current experimental facilities, but could be achieved in near future as proposed by Tajima [29].

8.3 Neutron Production with Laser-Accelerated Protons

We have performed experiments on two different laser systems with the aim to investigate the possibility of driving a pulsed neutron source by a laser. We have performed experiments on petawatt arm of the VULCAN Nd:Glass giant pulse laser at the Rutherford Appleton Laboratory, U.K., where we have measured proton acceleration and neutron generation through (p,xn) reactions on lead. However, because of their low repetition rate, giant pulse lasers are not perfectly suited for applications such as a pulsed neutron source driver. High-energy single-shot lasers can typically deliver pulses with repetition rates between 1 per every few hours and 1 per every 30 min.

Tabletop lasers typically operate at much higher repetition rate, which is measured in pulses per second. We have performed experiments on smaller and faster tabletop 10 TW Ti:Sapphire laser with 10 Hz repetition rate at the University of Jena, Germany. There we accelerated protons at lower intensities but with a possibility of a higher repetition rate. However, their energies were too small to induce nuclear reactions on lead and so we have not produced any neutrons [30].

We used a similar experimental setup at both laser facilities. The laser beam was focussed on the surface of thin foils (thickness around few tens of μm) of aluminum (Al) or titanium (Ti). Accelerated proton beams were found on both sides (front and rear) of the primary targets (see Fig. 8.2 for the general experimental layout).

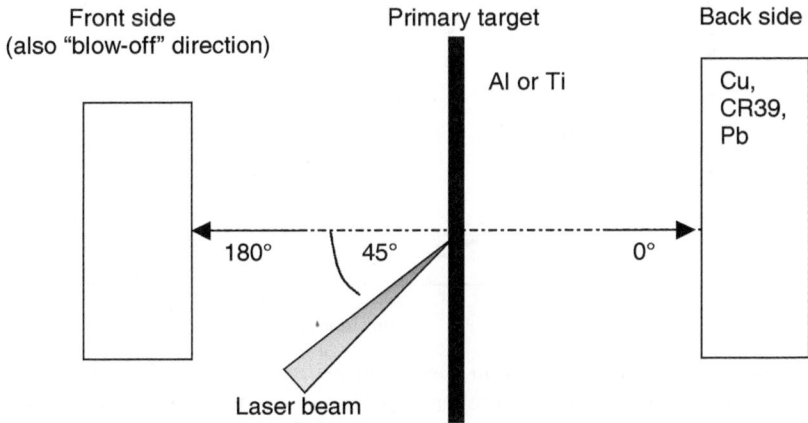

Fig. 8.2. General layout for proton acceleration experiments using high-intensity lasers. The laser beam was focussed on a thin primary target. Primary targets were aluminum and titanium foils in the VULCAN and Jena experiments, respectively. Protons accelerated from the back and front of the target were caught by thick secondary targets indicated with boxes on the schematics. The front side of the target is the side irradiated by the laser pulse. Secondary targets were copper (Cu) and lead (Pb) in the VULCAN experiments and CR39 plastic track detectors in the Jena experiments

The petawatt (1 PW = 10^{15} W) arm of the VULCAN Nd:glass laser was used to focus laser light with a wavelength of $\approx 1\,\mu m$ ($1\,\mu m = 10^{-6}\,m$) in a pulse of average duration 0.7 ps (1 ps = 10^{-12} s) onto a thin primary target (10-μm-thick Al foil). One laser pulse delivered 400 J of energy to a 7-μm-diameter focal spot (laser energy was 600 J before the compression). Peak laser intensity in the focus was 4×10^{20} W/cm^2. The typical broad energy spectrum of laser-accelerated protons was measured with copper stack exposition as described in [31, 32]. Two stacks of several 50 $\times$ 50-mm copper pieces, in total 3.7-mm thick, were positioned 38 mm from the primary target. One stack was positioned in the back and another in the front of the Al foil. Activity of the ^{63}Zn found in different layers of copper stacks was used to diagnose the energy distributions of protons, using known proton stopping powers in copper and well-known cross section for the ^{63}Cu(p,n)^{63}Zn reaction. The measured proton spectra accelerated from the front and the rear of the primary target are presented in Fig. 8.3. Both spectra show a Boltzmann-like distribution with significantly different temperatures and different high-energy cutoff values. A proton spectrum in front has a temperature of 2.5 MeV and maximum energy of 35 MeV, and the proton spectrum in the rear has a temperature of 4.2 MeV and maximum energy of 42 MeV. We calculated the total number of

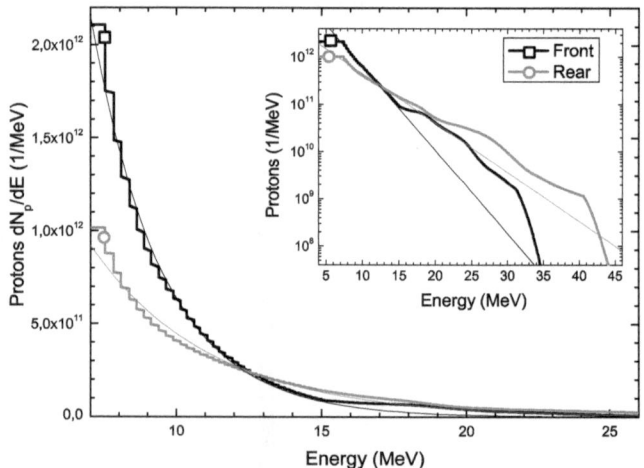

Fig. 8.3. Laser-accelerated proton spectra measured at the VULCAN facility. The energy of the protons and the number of protons per energy in a pulse were determined from the ^{63}Cu(p,n)^{63}Zn reactions in Cu foil stacks and from known proton stopping powers in copper (range of the protons). Two separate copper stacks were simultaneously positioned at the front and the rear of the laser-irradiated 10-μm-thick Al target. Fitted Boltzmann spectrums are presented for both cases with thin lines. Temperature of the protons on the front side was 2.5 MeV and the temperature on the back side was 4.2 MeV. The total number of accelerated protons to energies above 10 MeV was 7×10^{11} for the front side and 5×10^{11} for the back side

accelerated protons to energies above 10 MeV to be approximately 7×10^{11} for the front side and 5×10^{11} for the back side.

The Jena JETI (tabletop, 10 TW, 10 Hz) $(1 \, TW = 10^{12} \, W)$ Ti:Sapphire laser produces laser light with wavelength of 795 nm $(1 \, nm = 10^{-9} \, m)$ in pulses of average duration between 80 and 100 fs $(1 \, fs = 10^{-15} \, s)$. One laser pulse delivered $\approx 900 \, mJ$ of energy to a 4-µm-diameter focal spot on the surface of 5-µm-thin titanium target (laser energy was 1.2 J before the compression). Peak laser intensity in the focus was $2 \times 10^{19} \, W/cm^2$. The divergence and the maximum energy of laser-accelerated proton beams were measured with 50×50-mm CR39 track detectors positioned 34 mm behind the primary target [33]. CR39 plastic was covered with aluminum foils of different thicknesses (from 5 to 95 µm) to determine the maximum proton energy and beam divergence. Using different laser amplifier setups, we were able to measure proton acceleration characteristic at different laser intensities on target. Summary of these experiments is presented in Table 8.1. An example of scanned proton image on exposed detector is presented in Fig. 8.4. CR39 plastic is a commonly used solid state nuclear track detector (SSNTD). Because of its high sensitivity, it is especially appropriate for detection of charged particles of small and medium mass with low kinetic energy [34]. After the exposition, the tracks on the CR39 were visualized by chemical etching in the aqueous solution of NaOH. From the area on the track detector shaded by the visualized tracks, we were able to determine the number of aluminum foils penetrated by the protons and hence also the maximum kinetic energy of the protons in the beam. All areas seen as white on Fig. 8.4 have proton track density above the saturation level. From the known saturation density of proton tracks on CR39 under these etching conditions, we can calculate the minimum number

Table 8.1. The maximum energies of the protons accelerated at the Jena laser for three different target intensities. Stacks of aluminum foils and CR39 plastic were positioned 34 mm behind the laser-irradiated 5-µm-thick titanium target. The energy of the protons in a pulse was determined from the thickness of aluminum foil penetrated by the protons. One example of the VULCAN experiment data is presented in the last row for comparison

Laser Intensity (I_L) (W/cm^2)	$I_L \, \lambda^2$ $(W\mu m^2/cm^2)$	Maximum Al Thickness Penetrated (µm)	Maximum Proton Energy (MeV)
4×10^{18}	2×10^{18}	25	1.6 ± 0.1
10^{19}	6×10^{18}	65	2.7 ± 0.1
2×10^{19}	10^{19}	80	3.1 ± 0.1
[a]4×10^{20}	4×10^{20}	na	42

[a] VULCAN experiment is presented in this row. na = not applicable.

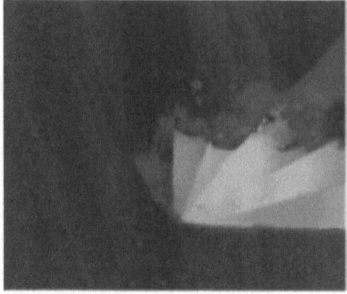

Fig. 8.4. An example of scanned proton image on exposed and etched CR39 plastic detector is presented on left, together with a photograph of aluminum filter used on right. Aluminum filter is made of 5-μm-thick Al foils (all together 19 pieces of different shapes). Aluminum thickness of the filter increases from 0 to 95 μm in contraclockwise direction. This particular detector was irradiated with laser irradiance of 2×10^{18} Wμm^2/cm^2. We can see clearly that protons have penetrated at least five segments of the aluminum filter (25 μm of aluminum)

of protons with kinetic energy above 3 MeV in the beam to be 10^8 1/sr per shot.

Proton energies produced in Jena were slightly smaller than expected from the data presented on Fig. 8.1 and from the extrapolation equation (8.1). However, with appropriate optimization of the experiments these proton energies could increase, as was shown by Mackinnon [35], up to the values predicted by equation (8.1). Even so, the maximum proton energies produced and detected in Jena at this laser irradiance are too low to induce (p,xn) nuclear reactions on high-Z materials as lead. However, neutrons can be produced through (p,xn) reactions on low-Z materials like beryllium, as will be shown later, even at these proton energies.

8.3.1 Proton-to-Neutron Conversion Through (p,xn) Reactions on Lead on VULCAN Laser

The proton pulse on VULCAN laser was converted into a neutron pulse with a 1-mm-thick, 5×5-cm natural lead sample positioned approximately 5 cm behind the primary target, along the target normal direction, subtending a solid angle of 1 sr. Several (p,xn) reactions on natural lead produce a variety of bismuth isotopes and several fast neutrons. The number of total neutrons produced was determined using two independent methods. In the first case, we calculated the number of neutrons from the known incoming proton spectrum. In the second case, we experimentally measured number of residual bismuth isotopes and hence also the number of neutrons released.

In the first case, the neutron spectrum (presented in Fig. 8.5) was calculated using the known incoming proton spectrum, proton stopping powers, and appropriate cross sections (presented in Fig. 8.6 [36]). We have taken

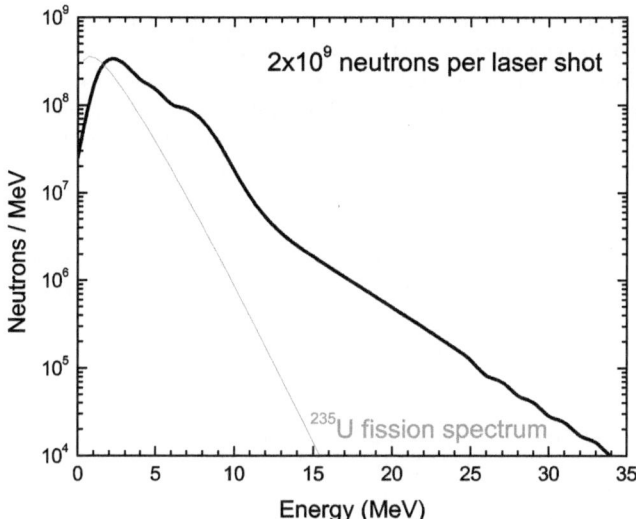

Fig. 8.5. Calculated neutron spectra released in ^{nat}Pb(p,xn)Bi reactions in VUL-CAN experiment. Cumulative number of neutrons per shot was determined also experimentally from the measured number of residual bismuth atoms in the lead target. Typical prompt ^{235}U fission spectrum is presented on the graph for comparison. Note that this ^{235}U spectrum was normalized to match the maximum of the laser-generated neutron spectrum

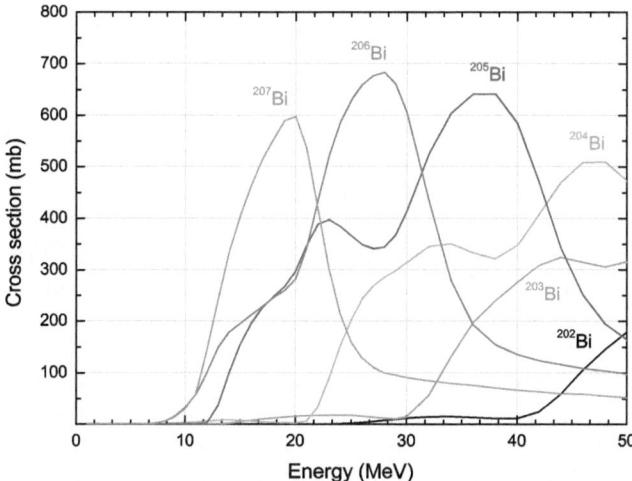

Fig. 8.6. Cross sections for the production of different bismuth isotopes from the natural mixture of lead isotopes taking into account (p,xn) and (p,γ) reactions (error bar on the cross section value is about 20%) [36]

Table 8.2. List of identified reaction products from proton-induced reactions on lead. For each of the nuclides listed, at least six of the main emission lines were identified. Since lead has several stable isotopes, there can be several (p,xn) reactions leading to the production of one residual bismuth isotope. Reactions marked with {} are listed only for completeness, since it is highly unlikely that these high threshold reactions occurred in our experiment. Because of their long half-lifes, isotopes ^{207}Bi and ^{208}Bi were not detected with γ spectroscopy

Isotope	Half-life	List of Possible Reaction Channels	Measured Number of Bismuth Atoms
^{202}Bi	1.67 h	^{204}Pb(p,3n), {^{206}Pb(p,5n), ^{207}Pb(p,6n), ^{208}Pb(p,7n)}	1.94×10^6 (1 ± 0.06)
^{203}Bi	11.76 h	^{204}Pb(p,2n), ^{206}Pb(p,4n), {^{207}Pb(p,5n), ^{208}Pb(p,6n)}	2.13×10^7 (1 ± 0.05)
^{204}Bi	11.22 h	^{204}Pb(p,n), ^{206}Pb(p,3n), ^{207}Pb(p,4n), {^{208}Pb(p,5n)}	6.42×10^7 (1 ± 0.05)
^{205}Bi	15.31 d	^{204}Pb(p,γ), ^{206}Pb(p,2n), ^{207}Pb(p,3n), ^{208}Pb(p,4n)	5.44×10^8 (1 ± 0.05)
^{206}Bi	6.24 d	^{206}Pb(p,n), ^{207}Pb(p,2n), ^{208}Pb(p,3n)	5.71×10^8 (1 ± 0.05)
^{207}Bi	31.57 y	^{206}Pb(p,γ), ^{207}Pb(p,n), ^{208}Pb(p,2n)	
^{208}Bi	3.7×10^5 y	^{207}Pb(p,γ), ^{208}Pb(p,n)	
^{209}Bi	stable	^{208}Pb(p,γ)	

into account only the most important (p,xn) reactions on ^{206}Pb, ^{207}Pb, and ^{208}Pb, which yield various bismuth isotopes (12 reactions in all as presented in Table 8.2). Note: natural lead consists of four isotopes: 1.4% ^{204}Pb, 24.1% ^{206}Pb, 22.1% ^{207}Pb, and 52.4% ^{208}Pb [37]. As we can see, the neutron spectrum has a peak around 2 MeV (see Fig. 8.5) and has a long but small tail to high energies, which extends beyond a typical uranium fission spectrum. From the calculated neutron spectrum, we see that the total number of neutrons released was in the order of 2×10^9 neutrons per laser shot.

In the second case, the number of produced γ-emitting residual nuclides was determined using γ-spectroscopy with a calibrated high-purity germanium detector (an example of the obtained spectra is given in [32]). After γ peak identification based on the measured γ energies and half-life, the net peak areas were used to calculate the numbers of each residual nuclide produced at the time of the laser shot (taking into account the detection efficiencies, decay branching ratios, gamma emission probabilities, and half-life). The main product nuclides were found to be close to the target nuclei, namely ^{206}Bi and ^{205}Bi, as can be seen from Table 8.2, where all bismuth reaction products are listed. In smaller qualities we have found also more "distant" isotopes of ^{204}Bi, ^{203}Bi, ^{202}Bi, and ^{203}Pb. Observed (p,3n) reaction on ^{204}Pb leading to ^{202}Bi is another experimental verification that the maximum proton energy was significantly higher than 21 MeV, as this reaction's threshold is $\approx$21.3 MeV. Higher energy reactions, for examples, the ^{204}Pb(p,4n)^{201}Bi with a reaction threshold of 28.8 MeV, were not observed. These reactions have also occurred but at a significantly lower reaction rate than the reactions mentioned before, and thus the product activity was below the detection limit. All together we have produced and measured more than 1.7×10^9 bismuth atoms. This value is in agreement with the predicted number of fast neutrons released in the (p,xn) reactions and so experimentally confirming the calculated value of 2×10^9 neutrons released per laser shot. Since we have not measured long-lived bismuth isotopes ^{207}Bi and ^{208}Bi, we may say that 2×10^9 neutrons is certainly conservative estimate for the total number of neutrons released. We should mention that this neutron pulse is certainly not isotropic but is forward directed because of momentum conservation of the forward-directed proton beam.

Natural lead is composed of four different isotopes, as described above. Protons will then react with all isotopes to induce (p,xn) or (p,γ) reactions to all of those. Therefore, the production of a specific bismuth isotope can result from different reactions on the different isotopes of natural lead. For example, ^{204}Bi, can be produced via the following reactions: ^{208}Pb(p,5n)^{204}Bi, ^{207}Pb(p,4n)^{204}Bi, ^{206}Pb(p,3n)^{204}Bi, and ^{204}Pb(p,n)^{204}Bi; similarly, ^{207}Bi could be a reaction product of the following three reactions: ^{208}Pb(p,2n)^{207}Bi, ^{207}Pb(p,n)^{207}Bi, and ^{206}Pb(p,γ)^{207}Bi. Using postirradiation gamma spectroscopy system to measure the bismuth production, we cannot distinguish the part of each production path without knowing the protons energy distribution exactly. On the other hand, individual cross sections for each reaction on each isotope of lead are known and can be combined to build the bismuth production cross section. Individual cross sections of different lead isotopes for each individual reaction are presented in Fig. 8.7. If we take these individual cross sections and add them together with proper weights according to the natural abundances of lead isotopes, we get the cross sections for the production of bismuth isotopes for natural lead as presented in Fig. 8.6.

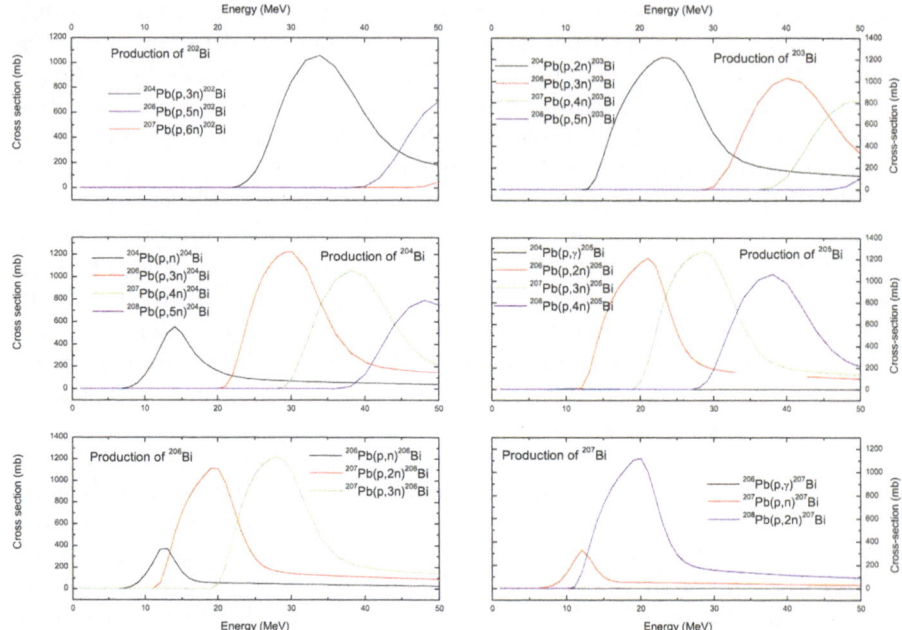

Fig. 8.7. Individual cross sections for the production of different bismuth isotopes. In each graph, all reaction channels leading to production of one bismuth isotope are shown [36]

8.4 Laser as a Neutron Source?

It is interesting to compare the above-mentioned laser-generated neutron source with existing traditional neutron sources. Traditional neutron sources can be, in general, divided into two main groups (Table 8.3). The first group consists of large neutron irradiation facilities such as reactors and proton accelerator-driven spallation sources producing high neutron fluxes. These installations are stationary, require a lot of trained personnel to operate and maintain them, and need very strict radiation shielding. The second group is smaller compact units, which can be, in turn, divided into two subgroups: passive neutron sources based on neutron emitting radioactive materials and small plasma-driven neutron generators. Neutron emitting radioactive materials can be either very high-Z materials undergoing spontaneous fission or mixtures of radioactive and additional low-Z materials emitting neutrons through (γ,n) or (α,n) reactions. These neutron sources cannot be switched off. Since they usually emit also other types of radiation, they must be heavily shielded and are very limited in strength. Plasma-driven neutron generators are based on deuterium or tritium fusion reactions [42]. They typically emit monoenergetic fusion neutrons but their lifetimes are usually very limited because of consumption of their deuterium (tritium) targets. Neutron generators based

Table 8.3. Overview of currently achievable neutron strengths for different commercially available neutron sources. Recent experimental results for laser-generated neutrons are added for comparison

Big Stationary Neutron Sources	
	Flux ($1/\mathrm{cm}^2\mathrm{s}$)
Traditional reactor	From 10^7 to 10^{13}
High flux research reactor	Up to $10^{15\mathrm{a}}$
Accelerator driven spallation	Up to $10^{14\mathrm{b}}$

Compact and Portable Neutron Sources	
	Typical Source Strength ($1/\mathrm{s}$)
Radioactive neutron sources[c]	10^5 to 10^7
Spontaneous fission sources[d]	Around 10^{10}
Portable neutron generators[e]	10^8 to 10^{10}

Lasers			
References	**Reaction(s) Used**	**Measured Source Strength (per shot)**	**Shot Energy (J)**
Hartke et al. [17]	D–D	2×10^3	0.2
Lancaster et al. [40]	$^7\mathrm{Li}(\mathrm{p,n})^7\mathrm{Be}$	2×10^8 1/sr	69
Yang et al. [41]	$^{\mathrm{nat}}\mathrm{Zn}(\mathrm{p},x\mathrm{n})\mathrm{Ga}$	$\approx 10^{10}$	230
Yang et al. [41]	$^7\mathrm{Li}(\mathrm{p,n})^7\mathrm{Be}$	$5 \times 10^{10\mathrm{f}}$	230
This work	$^{\mathrm{nat}}\mathrm{Pb}(\mathrm{p},x\mathrm{n})\mathrm{Bi}$	2×10^9	400

[a] Modern high flux reactors like FRM-II at TU-München [38] can reach such high thermal fluxes.
[b] SINQ at Paul Scherrer Institut [39] is the world's most powerful spallation neutron sources today.
[c] Neutron sources using (α,n) reactions (e.g., $^{226}\mathrm{Ra}$-α-Be) or radioactive photoneutron sources (e.g., $^{54}\mathrm{Mn}$-γ-D).
[d] Spontaneous fission sources are usually small (e.g., few mg) and they emit typical fission spectrum neutrons at high specific rate (e.g., 2.3×10^{12} per gram of $^{252}\mathrm{Cf}$).
[e] Portable neutron generators based on D–D, D–T, or T–T reactions using RF heated plasmas as a source for fast ions.
[f] This value was calculated only with Monte Carlo simulation and was not experimentally verified.

on D–D or even D–T fusion reactions can be driven with small electrostatic accelerators and could be shaped into portable units [43]. Compact neutron sources deliver smaller fluxes than large installations, but are also easier to maintain. Since compact units normally emit neutrons isotropically, the neutron flux from these sources is decreasing with the square of distance from the source. We mention also possibility of producing neutrons with electron accelerators for completeness. The kinetic energy of the electrons is lost as bremsstrahlung. These bremsstrahlung photons interact with nuclei to produce neutrons in (γ,n) reactions [42].

Laser-produced neutrons can be divided into two general groups. First group is neutrons produced in deuterium or tritium cluster fusion reactions. Such cluster fusion sources are capable of yields up to 10^5 fusion neutrons per Joule of incident laser energy using pure deuterium clusters [17]. One example of such neutron source is presented also in Table 8.3. Another separate group of neutron sources are neutron sources discussed here in this work. However, we need significantly higher laser intensities to generate neutrons, using accelerated protons in comparison to deuterium fusion neutron generation. Laser intensities around $10^{17}\,\mathrm{W/cm^2}$ are enough to generate, neutrons with D–D fusion in contrast to around $10^{19}\,\mathrm{W/cm^2}$ needed to start the research on (p,n) reactions. On the other hand, the neutron yields are much bigger in this second group, as we can see (Table 8.3) yields 10^7 neutrons per Joule of incident laser energy can be achieved.

According to our experiments and values quoted in recent publications (see also Table 8.3), current giant pulse laser systems (e.g., VULCAN) can produce approximately 10^{10} neutrons per shot. This value is small compared to large-scale neutron sources, but they can be directly compared to strengths of compact neutron sources, even if we take into account the low repetition rate of current giant pulse laser systems (e.g., one shot per few hours). In addition, laser-generated neutron sources are not isotropic and hence neutron flux from such sources is stronger in forward direction. This directionality could be very useful for selected applications (e.g., neutron radiography, BNCT, ...). When comparing the laser-generated neutron sources, with traditional neutron sources, we must also mention extremely short-pulse durations achievable by lasers. Currently, pulse durations less than nanosecond can easily be achieved with lasers.

8.5 Optimization of Neutron Source – Nuclear Applications with Future Laser Systems?

We have demonstrated the possibility of generating neutron pulses, but their intensities might be slightly under the limits needed for useful applications of this novel neutron source. Of course, the strength of any, not only laser neutron source, could be increased by surrounding it with a thin layer of fissile material – this is known as neutron booster concept [44]. Typical, increases by a factor of 10 were demonstrated in calculations, taking into account multiplication of neutrons in fissile material. But, it is preferable to study the optimization of laser neutron sources without introducing any fissile material, which is inherently connected to radiation protection and criticality control problems. We can show that anticipated improvement in high-intensity laser technology will lead also to higher laser-generated neutron strengths. Both proton-to-neutron conversion efficiency and laser repetition rates are low at the present time and will increase with laser development. In our experiment we observed a small proton-to-neutron conversion efficiency. From the total number of protons with energies above 10 MeV (5×10^{11}) that have entered the lead sample, 2×10^9 neutrons were produced giving a proton-to-neutron conversion efficiency ε_{pn} in the order of 4×10^{-3} for a laser irradiance of 4×10^{20} Wμm^2/cm^2 and a laser energy on target of 400 J.

8.5.1 Laser Light-to-Proton and Proton-to-Neutron Conversion Efficiencies

Neutron production efficiency is a product of laser light-to-proton efficiency (ε_{Lp}) and proton-to-neutron conversion efficiency (ε_{pn}). As was shown above, ε_{pn} is small at current laser intensities. This fact can be easily understood if we look into the basic processes behind the conversion. To generate neutrons, protons must interact with the target nucleus via a neutron-generating nuclear reaction. The probability for the nuclear reaction can be described with the mean free path (Λ) parameter defined as

$$\Lambda = \frac{1}{\Sigma} = \frac{1}{N\sigma} , \qquad (8.2)$$

where Σ is the macroscopic and σ microscopic cross section of the reaction in question and N is the atom density of the material. For neutrons, which travel at constant velocity between collisions, Λ measures the average distance a neutron is likely to travel before colliding with a nucleus in a particular nuclear reaction. However, charged particles like protons experience continuous energy loss passing through matter and hence have a limited range (R) in solids. For protons with energies below a few hundred MeV R are a few orders of magnitude smaller than any Λ for proton-induced nuclear reactions. In this energy range, where Λ is few orders of magnitude larger than R, we can introduce a ratio between both parameters as a simplistic measure to estimate the proton-to-neutron conversion efficiency as

$$\varepsilon_{pn} \approx \frac{R}{\Lambda} . \qquad (8.3)$$

Calculated ε_{pn} for lithium, lead and uranium are presented in Fig. 8.8 together with the Λ and R used.

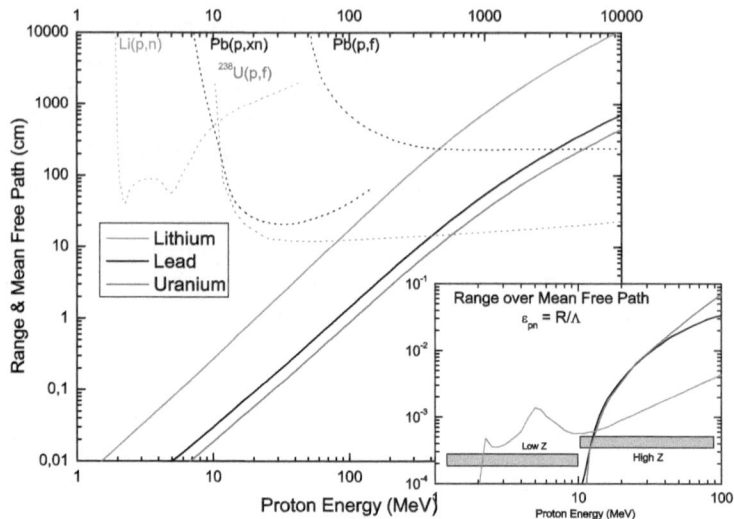

Fig. 8.8. Proton ranges (R, *solid lines*) in three different metals together with mean free paths (Λ, *dotted lines*) for neutron generation reactions in the same materials as a function of incoming proton energy. In the small insert the ratio between R and Λ as a function of proton energy is shown. This ratio is directly related to the efficiency of proton-to-neutron conversion ε_{pn} in different materials. Li and Pb are two typical materials used for neutron production, and we have included U as a typical fissionable actinide material for which proton transmutation is also very interesting. The proton ranges were calculated with SRIM-2003 [45] and evaluated cross-sectional data for proton-induced reactions were retrieved from IAEA-NDS Database on Experimental Nuclear Reaction Data (EXFOR) [46]

It can be seen that for protons at 5 MeV in Li the range is approximately 1 mm and the Λ for Li(p,n) reaction is 50 cm. So we can estimate that less than 1 in 500 protons will induce (p,n) reactions or ε_{pn} is smaller than 0.002. We can see that ε_{pn} in lithium is higher than ε_{pn} in lead for protons below 10 MeV. But it is much higher in lead at proton energies above 10 MeV. This fact, which is already well known from proton accelerator-driven neutron source studies, suggests that we need higher proton energies for efficient neutron generation. For our experiment described in this chapter we can see that the ε_{pn} for lead is between 10^{-3} and 10^{-2} for proton energies between 20 and 40 MeV. Taking into account that proton energies were between 20 and 40 MeV in our experiment, this value is in a perfect agreement with the measured conversion efficiency of 4×10^{-3} if we take into account the simplicity of the approach used.

The question of the laser light-to-proton conversion efficiency is more specific and has not yet been properly answered. Most of the published studies on laser acceleration of protons were focussed only on the question of maximum proton energy reachable. Few studies on this subject [47] indicate a correlation between ε_{Lp}, the laser pulse energy E, and the laser irradiance $I\lambda^2$. For a laser pulse energy 100 J and laser irradiance 10^{20} Wµm²/cm², the ε_{Lp} is around 10%. For smaller E, and $I\lambda^2$ the ε_{Lp} is also smaller. For $E \approx 1$ J and $I\lambda^2 \approx 10^{18}$ Wµm²/cm², an ε_{Lp} as low as 0.001% can be found in reports [47]. However, large-experiment-to experiment variations in the data can be found which are related to different definitions of ε_{Lp}, different measurement techniques, laser contrasts, and pulse durations used in the different experiments. In addition, ε_{Lp} depends also on primary target thicknesses and can be increased by target surface contamination control.

8.5.2 High-Intensity Laser Development

State-of-the-art tabletop laser systems operate at 10 Hz repetition rate and deliver approximately 1 J of light energy to a target with a focal irradiance of 10^{19} Wµm²/cm². They can accelerate protons to a few MeV, which are capable of inducing (p,n) reactions in lithium. The efficiency ε_{Lp} for such lasers is at least 10^{-5} [47] and the proton-to-neutron conversion efficiency ε_{pn} for protons with energies between 1 and 3 MeV is 5×10^{-4}. Such a tabletop laser system with a thick Li target can thus produce more than 10^5 neutrons per shot or 10^6 neutrons per second. This neutron strength is comparable to available californium or Ra–Ba neutron sources and is an order of magnitude stronger than neutron sources generated with interaction of femtosecond laser pulses with deuterium clusters [17, 48].

Because of the relatively high cross section for the (p,f) reaction in uranium and the (p,xn) reaction in lead the ε_{pn} will reach approximately 1% for these reactions, using laser-accelerated protons with maximum energies around 100 MeV. According to the relation between measured proton energies and laser irradiance $I\lambda^2$ (see equation 1), 100 MeV protons will be reached at 10^{21} Wµm²/cm². This laser irradiance is currently just over the available limit of giant pulse lasers, and Yang has showed [41] that at least 5×10^{10} neutrons per shot can be expected at these laser irradiances.

This level of laser intensity is likely to be achievable also on diode-pumped tabletop lasers systems in the near future. POLARIS [49, 50] – a diode-pumped high-power laser system under construction at the University of Jena – will deliver ultrashort pulses (150 fs) with a planned energy up to 200 J and a wavelength of 1030 nm at a predicted repetition rate of 0.1 Hz due for completion in 2007. Focusing this light to spots with diameters of around 10 µm will result in an irradiance of 10^{21} Wµm²/cm². POLARIS will thus be able to support a compact neutron source of at least 5×10^{10} neutrons per pulse, that is, 5×10^9 neutrons per second. Even if we do not take into account the increase of ε_{Lp} and assume constant efficiency even at higher laser irradiance and pulse energy, we can assume an increase in the neutron source strength due to an increase in the proton-to-neutron conversion efficiency ε_{pn}. We conclude that petawatt tabletop lasers will be able to deliver pulsed neutron sources with 10^{11} neutrons per pulse in nanosecond pulses. The neutron activation capability of such a laser source is comparable to a neutron source with a continuous strength of 10^{10} neutrons per second. Such neutron flux levels will require shielding, activity, and dosimetry control infrastructures similar to a small nuclear facility. This neutron source strength could provide neutron flux on the order of $10^{12}\,1/\mathrm{cm}^2\mathrm{s}$ for small samples with dimensions on the order of $1\,\mathrm{mm}$. Such neutron source strength is sufficient for several neutron applications in the fields of nuclear geophysics and neutron radiography. Extremely short neutron pulse lengths make this neutron source very interesting for ultrafast neutron radiography and ultrafast neutron activation analysis.

With the currently fast evolution of laser technology, laser systems with higher repetition rates and higher intensities are emerging. A tabletop laser with a target intensity of $10^{21}\,\mathrm{W/cm}^2$ and a repetition rate of $100\,\mathrm{Hz}$ could theoretically produce a neutron source of strength up to 10^{13} neutrons per second.

Some authors [29] have indicated that future laser systems will reach laser intensities well beyond $10^{21}\,\mathrm{W/cm}^2$, even up to $10^{24}\,\mathrm{W/cm}^2$, in the next decade. We can expect to see even stronger neutron sources in these laser operating regimes; however, our analysis is not necessarily directly transferable. At $10^{22}\,\mathrm{W/cm}^2$ the laser intensities will be high enough to accelerate protons up to energies of $350\,\mathrm{MeV}$. The length of the mean free path for proton-induced fission in uranium is comparable to the proton range at this energy, resulting in an extremely high efficiency of laser-induced fissions. At these proton energies spallation reactions start to dominate in high-Z solid targets and they are even more efficient in neutron production than fission reactions. All this development will make laser-generated neutron sources even more efficient, and we can speculate that they will be able to produce more than 10^{13} neutrons per second. On the other hand, laser target areas for such high-intensity laser systems will require heavy shielding against γ and fast neutron radiation, making them similar to nuclear facilities.

8.6 Conclusions

The highly efficient conversion of laser light into a fast proton beam, achieved with irradiation of thin solid targets, has opened up a possibility to generate pulsed neutron sources in a completely new way. These sources are based on proton-to-neutron conversion in thick converter materials. We showed that low-Z materials like lithium can be used as proton-to-neutron converters in current laser systems; however, high-Z materials, such as lead, are the materials of choice for efficient proton-to-neutron conversion in near future laser systems, when higher proton energies will be available. These neutron sources have a forward-peaked neutron flux, with a continuous energy spectrum and an extremely short pulse width.

We have experimentally demonstrated the production of 2×10^9 neutrons per laser shot on the VULCAN laser, using proton-to-neutron conversion in lead. Using proton-to-neutron conversion in lithium, current state-of-the-art tabletop lasers can theoretically generate neutron pulses at a rate between 10^6 and 10^7 neutrons per second. This is an order of magnitude stronger than neutron sources generated with the interaction of femtosecond laser pulses with deuterium clusters. It was also demonstrated that, with tabletop lasers under construction, pulsed neutron sources with 10^{10} fast neutrons per second in pulses smaller than 1 ns will be achievable. Extremely short neutron pulse lengths make this neutron source very interesting for ultrafast neutron activation or ultrafast neutron irradiation material damage studies. For applications, which need thermalized or a moderated neutron spectrum, this neutron source might also be applicable, but fluxes for collimated thermalized neutrons will be a few orders of magnitude smaller. However, with the fast evolution of laser technology, we can speculate to see laser systems that could support pulsed neutron sources at a continuous strength equal to 10^{13} neutrons per second. Taking all this into account, a fast, cheap, flexible, and pulsed neutron source could be imagined without resource to nuclear reactors or proton accelerators. Even with laser intensities reachable today, laser neutron sources can be compared to available californium (spontaneous fission) or Ra–Ba neutron sources.

References

1. D. Umstadter: Nature **404**, 239 (2000)
2. K. Ledingham, P. McKenna, R.P. Singhal: Science **300**, 1107 (2003)
3. J. Galy et al.: Central Laser Facility Annual Report 2001/2002, 29, (2002) http://www.clf.rl.ac.uk/Reports/
4. F. Ewald et al.: Plasma Phys. Control. Fusion **45**, A83 (2003)
5. J. Magill, H. Schwoerer, F. Ewald, J. Galy, R. Schenkel, R. Sauerbrey: Appl. Phys. B **77**, 387 (2003)
6. H. Schwoerer, F. Ewald, R. Sauerbrey, J. Galy, J. Magill, V. Rondinella, R. Schenkel, T. Butz: Europhys. Lett. **61**, 47 (2003)
7. B. Liesfeld et al.: Appl. Phys. B **79**, 1047 (2004)
8. S. Karsch et al.: Phys. Rev. Lett., **91**, 015001 (2003)
9. K. Ledingham et al.: Phys. Rev. Lett., **84**, 899 (2000)
10. P. Main et al.: IEEE J. Quant. Electr., **24**, 398 (1988)
11. RAL, Chilton, Didcot, UK, http://www.clf.rl.ac.uk/Reports/
12. LLNL, Livermore, CA, http://www.llnl.gov/
13. LULI, Palaiseau, France, http://www.luli.polytechnique.fr/
14. CUOS, Ann Arbor, MC, http://www.eecs.umich.edu/USL/
15. LOA, Palaiseau, France, http://wwwy.ensta.fr/loa/
16. Y. Sentoku et al.: Phys. Plasmas **10**, 2009 (2003)
17. R. Hartke, D.R. Symes et al.: Nucl. Instrum. Methods Phys. Res. A **540**, 464 (2005)

18. S.J. Parry: *Activation Spectrometry in Chemical Analysis* (John Wiley and Sons, New York, 1991)
19. M.D. Glascock: University of Missouri Research Reactor (MURR), Columbia, An Overview of Neutron Activation Analysis (2005) `http://www.missouri.edu/≈glascock/naa_over.htm`
20. Nuclear Geophysics and Its Applications. IAEA Technical Reports Series **393**, IAEA, Vienna, Austria (1999)
21. J.C. Domanus: *Practical Neutron Radiography* (Kluwer Academic Publishers, 1992)
22. E. Lehmann: *What Is Neutron Radiography?* (Paul Scherrer Institute, Villigen, Switzerland) `http://neutra.web.psi.ch/What/index.html`
23. J.T. Mendonça et al.: Meas. Sci. Technol. **12**, 1801 (2001)
24. R.A. Snavely et al.: Phys. Rev. Lett. **85**, 2945 (2000)
25. M. Kaluza et al.: Phys. Rev. Lett. **93**, 045003 (2004)
26. E.L. Clark et al.: Phys. Rev. Lett. **85**, 1654 (2000)
27. I. Spencer et al.: Nucl. Instrum. Methods Phys. Res. B **183**, 449 (2001)
28. T. Esirkepov et al.: Phys. Rev. Lett. **92**, 175003 (2004)
29. T. Tajima, C. Mourou: Phys. Rev. Spec. Topics **5**, 031301 (2002)
30. T. Žagar, J. Galy, J. Magill and M. Kellett: N. J. Phys. **7**, 253 (2005)
31. J.M. Yang, P. McKenna et al.: Appl. Phys. Lett. **84**, 675 (2004)
32. P. McKenna et al.: Phys. Rev. Lett. **94**, 084801 (2005)
33. T. Žagar et al.: Characterization of Laser Accelerated Protons with CR39 Track Detectors: Jena August 2004. S.P./K.04.224, EC-JRC-ITU, Karlsruhe (2004)
34. R. Ilić, S.A. Durrani: Solid state nuclear track detectors In: *M. F. L'Annunziata, Handbook of Radioactivity Analysis*, 2nd edn (Academic Press, Amsterdam, 2003), pp. 179–237
35. A.J. Mackinnon et al.: Phys. Rev. Lett. **88**, 215006 (2002)
36. A.J. Koning, S. Hilaire, and M.C. Duijvestijn: AIP Conf. Proc. **769**, 1154 (2005)
37. J. Magill: Nuclides.net. Springer-Verlag, Berlin (2002) `http://www.nuclides.net/`
38. Forschungsneutronenquelle Heinz Maier-Leibnitz (FRM II), TU-München, Garching, Deutchland. `http://www.frm2.tum.de/`
39. Paul Scherrer Institut, Villigen, Schweiz. `http://www.psi.ch/`
40. K. Lancaster et al.: Phys. Plasmas **11**, 3404 (2004)
41. J. Yang et al.: J. Appl. Phys. **96**, 6912 (2004)
42. J. Byrne: *Neutrons, Nuclei and Matter, an Exploration of the Physics of Slow Neutrons* (Institute of Physics, London, 1995)
43. Portable Neutron Generators, Del Mar Ventures, San Diego, CA. `http://www.sciner.com/Neutron/Neutron_Generators_Basics.htm`
44. J. Galy et al.: Nucl. Instrum. Methods Phys. Res. A **485**, 739 (2002)
45. J. Ziegler, J. Biersack: SRIM-2003: The Stopping and Range of Ions in Matter (2003) `http://www.srim.org/`
46. International Atomic Energy Agency – Nuclear Data Section, Vienna, Austria (2004) `http://www-nds.iaea.org/`
47. P. McKenna et al.: Rev. Sci. Instrum. **73**, 4176 (2002)
48. J. Zweiback et al.: Phys. Rev. Lett. **85**, 3640 (2000)
49. R. Sauerbrey et al.: POLARIS – a Compact, Diode-Pumped Laser System in the Petawatt Regime. International Workshop Lasers & Nuclei, Karlsruhe, 13–15 September 2004 S.P./K.04.173, EC-JRC-ITU, Karlsruhe (2004)
50. J. Hein et al.: POLARIS: An All Diode-Pumped Ultrahigh Peak Power Laser for High Repetition Rate. *Lasers and Nuclei* (Springer-Verlag, Berlin, in press)

Part III

Transmutation

9

Laser Transmutation of Nuclear Materials*

J. Magill[1], J. Galy[1], and T. Žagar[2]

[1]European Commission, Joint Research Centre, Institute for Transuranium Elements, Postfach 2340, 76125 Karlsruhe, Germany
Joseph.Magill@cec.eu.int
[2]Institute Jozef Stefan, Reacto Physics Department, Jamova 39, 1000 Ljubljana, Slovenia
tomaz.zagar@ijs.si

9.1 Introduction

Following the discovery of radioactivity by Becquerel in 1896, two young scientists, Frederic Soddy and Ernest Rutherford then at McGill University in Canada, set about to investigate the recently discovered phenomena. In 1901, the 24-year-old chemist Soddy (Fig. 9.1) and Rutherford were attempting to identify a gas that was being released from samples of radioactive thorium oxide. They believed that this gas – they called it an "emanation"– was related to the radioactivity of the thorium sample. In order to investigate the nature of this gas, Soddy passed it over a series of powerful chemical reagents, heated white-hot. No reactions took place. Years later in his biography, he wrote [1],

> I remember quite well standing there transfixed as though stunned by the colossal import of the thing and blurting out – or so it seemed at the time: "Rutherford, this is transmutation: the thorium is disintegrating and transmuting itself into argon gas." Rutherford's reply was typically aware of more practical implications, "For Mike's sake, Soddy, don't call it transmutation. They'll have our heads off as alchemists."

Following this discovery, Rutherford and Soddy published nine joint papers between 1902 and 1903 in a period of extremely productive research [2]. In 1902, they described their theory of radioactivity as a spontaneous disintegration of the radioactive element by the expulsion of particles with the result that new elements are formed. This was the ultimate step on the ancient alchemists' dream of transmutation (Fig. 9.2).

---

*Sections of this chapter have been taken from J. Magill, J. Galy: Radioactivity, Radionuclides, Radiation. (Springer-Verlag, 2005)

J. Magill et al.: *Laser Transmutation of Nuclear Materials*, Lect. Notes Phys. **694**, 131–146 (2006)
www.springerlink.com © Springer-Verlag Berlin Heidelberg and European Communities 2006

Fig. 9.1. Frederic Soddy (1877–1956) with permission from The Nobel Foundation

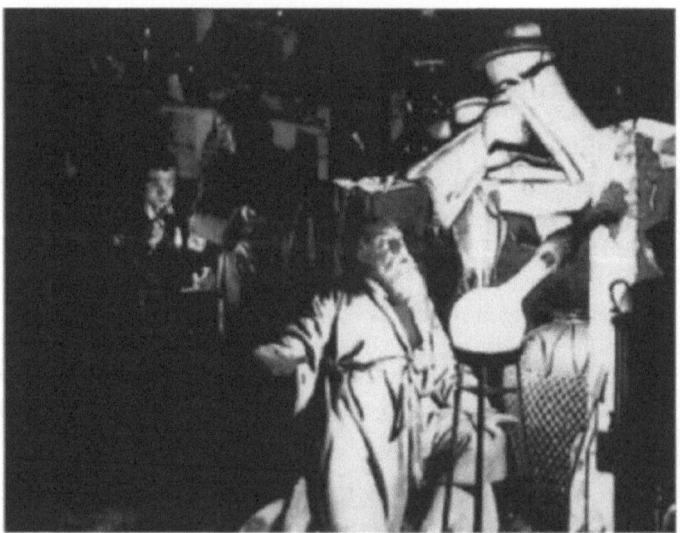

Fig. 9.2. "The Alchymists in Search of the Philosopher's Stone," Joseph Wright (1734–1797) with permission from Derby Museum and Art Gallery

Transmutation – the idea of changing one element into another – is almost as old as time itself. In the Middle Ages, the alchemists tried to turn base metals into gold. Transmutation, however, became only a reality in the last century with the advent of nuclear reactors and particle accelerators.

One of the main interests today in transmutation is in the field of nuclear waste disposal. Nuclear waste is a radioactive by-product formed during normal reactor operation through the interaction of neutrons with the fuel and container materials. Some of these radioactive by-products are very long-lived (long half-lifes) and, if untreated, must be isolated from the biosphere for very long times in underground repositories. Various concepts are being investigated worldwide on how to separate out (partition) these long-lived by-products from the waste and convert (transmute) them into shorter lived products, thereby reducing the times during which the waste must be isolated from the biosphere.

In the following sections, a brief history of attempts made to modify the decay constant, and thereby enhance the transmutation rate, is outlined. Thereafter, a new technique – laser transmutation – is described in which very high intensity laser radiation is used to produce high-energy photons, and particles that can be used for transmutation studies. Finally, some potential applications for "Homeland Security" are mentioned.

9.2 How Constant Is the Decay Constant?

Following the discovery of radioactivity, many attempts to modify α decay rates were made by changing temperature, pressure, magnetic fields, and gravitational fields (experiments in mines and on the top of mountains, using centrifuges). In one attempt, Rutherford [3] actually used a bomb to produce temperatures of $2,500°C$ and pressures of $1,000\,bar$ albeit for a short period of time. No effect on the decay constant was detected.

Only through the Gamow theory of alpha decay could one understand why the above experiments to modify the decay constant were negative. Gamow showed that quantum mechanical tunneling through the Coulomb barrier was responsible for alpha emission. Even if the entire electron cloud surrounding a nucleus were removed, this would change the potential barrier by only a very small factor. Changes of the order of $\delta k/k \approx 10^{-7}$ are to be expected, where k is the decay constant.

In 1947, Segre [4] suggested that the decay constant of atoms undergoing electron capture (EC) could be modified by using different chemical compounds of the substance. Different compounds will have different electron configurations and this should lead to small differences in the EC decay rate. This idea was confirmed experimentally using ^7Be. This nuclide has a half-life of 53.3 d and decay by EC is accompanied by the emission of a 477.6 keV gamma photon. A comparison of BeF_2 and Be revealed a difference in the decay rate $\delta k/k = 7 \times 10^{-4}$. These chemically induced changes in the decay constant are small but measurable. It is also to be expected that the decay constant can be modified by pressure.

As the pressure increases, the electron density near the nucleus should increase and manifest itself in an increase in the decay rate (for EC). Experiments [5, 6] on ^{99m}Tc, ^7Be, ^{131}Ba, and ^{90m}Nb have shown that this is indeed the case. The fractional change in the decay constant is $\delta k/k \approx 10^{-8}$ per bar. At pressures of 100 kbar, which can be relatively easily produced in laboratory conditions, $\delta k/k = 10^{-3}$ and the change in the decay constant is still small. Extrapolation to very high pressures would give $\delta k/k \approx 10$ at 1 Gbar and $\delta k/k \approx 10^3$ at 100 Gbar. With regard to β decay, it is also expected that screening effects can also modify the decay constant [7, 8]. Recently fissioning of ^{238}U has been demonstrated using very high power laser radiation (see later). The fact that through laser-induced fission one can significantly alter the rate of fission is however not achieved through modifying the environment. It arises indirectly through bremsstrahlung and electron-induced reactions with the nucleus. In the focal region, the beam diameter is $\cong 1\,\mu m$ and the penetration depth is 20 nm. In this region there are approximately 10^9 atoms of ^{238}U. On average, every 10 y one of these atoms will decay by alpha emission. Spontaneous fission will occur on a time scale approximately six orders of magnitude longer, that is, 10^7 years. Under irradiation by the laser, typically 8,000 fissions are produced per pulse.

9.3 Laser Transmutation

Recent advances in laser technology now make it possible to induce nuclear reactions with light beams [9, 10, 11]. When focussed to an area of a few tens of square microns, the laser radiation can reach intensities greater than $10^{20}\,\mathrm{W/cm^2}$. By focussing such a laser onto a target, the beam generates a plasma with temperatures of ten billion degrees $(10^{10}\,\mathrm{K})$ – comparable to those that occurred one second after the "big bang."

With the help of modern compact high-intensity lasers (Fig. 9.3), it is now possible to produce highly relativistic plasma in which nuclear reactions such as fusion, photo-nuclear reactions, and fission of nuclei have been demonstrated to occur. Two decades ago, such reactions induced by a laser beam were believed to be impossible. This new development opens the path to a variety of highly interesting applications, the realization of which requires continued investigation of fundamental processes by both theory and experiment and in parallel the study of selected applications. The possibility of accelerating electrons in focussed laser fields was first discussed by Feldman and Chiao [12] in 1971. The mechanism of the interaction of charged particles in intense

Fig. 9.3. *Left:* Giant pulse VULCAN laser. Courtesy: CCLRC Rutherford Appleton Laboratory. *Right:* High-intenisty Jena tabletop laser JETI. Courtesy: Institut für Optik und Quantenelektronik, Friedrich-Schiller-Universität, Jena

electromagnetic fields, for example, in the solar corona, had, however, been considered much earlier in astrophysics as the origin of cosmic rays. In this early work, it was shown that in a single pass across the diffraction limited focus of a laser power of 10^{12} W, the electron could gain 30 MeV, and become relativistic within an optical cycle. With a very high transverse velocity, the magnetic field of the wave bends the particle trajectory through $\boldsymbol{v} \times \boldsymbol{B}$ Lorentz force into the direction of the travelling wave. In very large fields, the particle velocity approaches the speed of light and the electron will tend to travel with the wave, gaining energy as it does so.

Dramatic improvements in laser technology since 1984 (Fig. 9.4) have revolutionized high-power laser technology [13]. Application of chirped pulse amplification techniques [14, 15] has resulted laser intensities in excess of 10^{19} W/cm^2. In 1985, Rhodes et al. [16] discussed the possibility of laser intensities of $\approx 10^{21}$ W/cm^2, using a pulse length of 0.1 ps and 1 J of energy. At this intensity, the electric field is 10^{14} V/cm a value which is over 100 times the coulomb field binding atomic electrons. In this field, a uranium atom will lose 82 electrons in the short duration of the pulse. The resulting energy density of the pulse is comparable to a 10 keV blackbody (equivalent light pressure ≈ 300 Gbar) and comparable to thermonuclear conditions (thermonuclear ignition in DT occurs at about 4 keV).

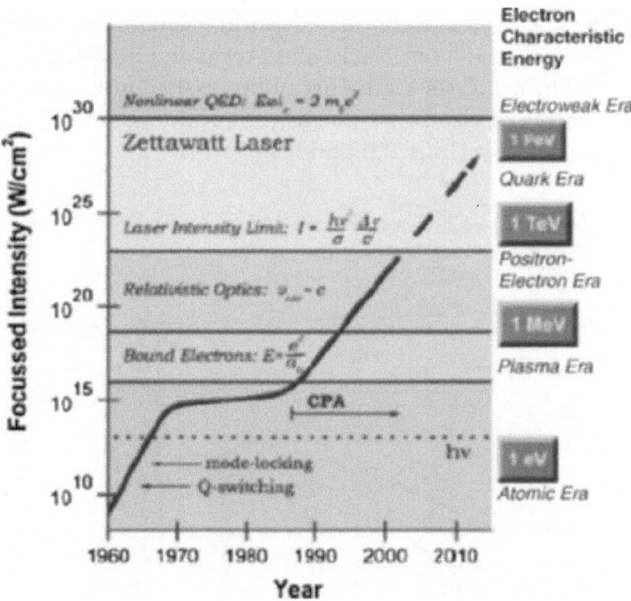

Fig. 9.4. Dramatic increase in focussed laser intensity over the past few decades for tabletop systems [13]. With the development of chirped pulse amplification (CPA) techniques in the mid-eighties, a new era of laser–matter interactions has become possible

In 1988, Boyer et al. [17] investigated the possibility that such laser beams could be focussed onto solid surfaces and cause nuclear transitions. In particular, irradiation of a uranium target could induce electro- and photo-fission in the focal region. These developments open the possibility of "switching" nuclear reactions on and off by high-intensity ultraviolet laser radiation and providing a bright point source of fission products and neutrons.

9.3.1 Laser-Induced Radioactivity

When a laser pulse of intensity 10^{19} W/cm^2 interacts with solid targets, electrons of energies of some tens of MeV are produced. In a tantalum target, the electrons generate an intense highly directional γ-ray beam that can be used to carry out photo-nuclear reactions. The isotopes ^{11}C, ^{38}K, ^{62,64}Cu, ^{63}Zn, ^{106}Ag, ^{140}Pr, and ^{180}Ta have been produced by (γ,n) reactions, using the VULCAN laser beam.

9.3.2 Laser-Induced Photo-Fission of Actinides – Uranium and Thorium

The first demonstrations were made with the giant pulse VULCAN laser in the United Kingdom, using uranium metal and with the high repetition rate laser at the University of Jena with thorium samples (experimental setup shown in Fig. 9.5). Both experiments were carried out in collaboration with the Institute for Transuranium Elements in Karlsruhe. Actinide photo-fission was achieved in both U and Th, using the high-energy bremsstrahlung radiation produced by laser acceleration of electrons. The fission products were identified by time-resolved γ-spectroscopy (Figs. 9.6 and 9.7).

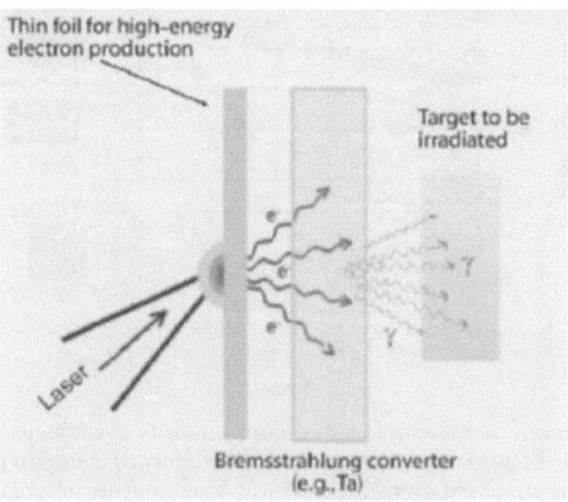

Fig. 9.5. Schematic setup of the laser experiments

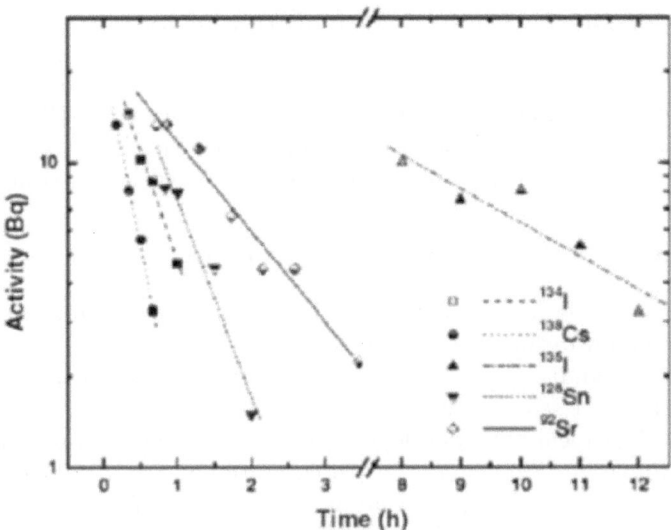

Fig. 9.6. Decay characteristics of fission products from bremsstrahlung-induced fission of ^{232}Th. The deduced half-lives are in good agreement with literature values. Symbols indicate experimental data [11]

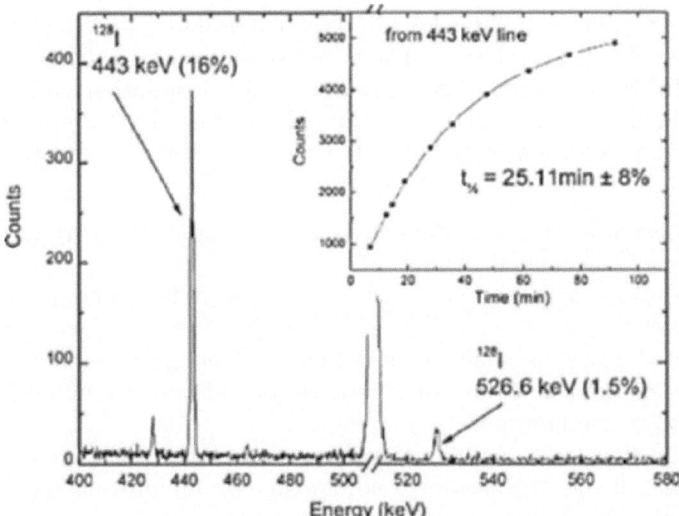

Fig. 9.7. Gamma emission spectra from one of the iodine samples measured before and after laser irradiation of the gold target. Characteristic emission lines of ^{128}I at 443.3 and 527.1 keV are clearly observed, alongside peaks from the decay of ^{125}Sb impurity and a peak at 511 keV from positron annihilation [18, 19]

9.3.3 Laser-Driven Photo-Transmutation of Iodine-129

The first successful laser-induced transmutation of ^{129}I, one of the key radionuclides in the nuclear fuel cycle was reported recently [18, 19, 20]. ^{129}I with a half-life of 15.7 million years is transmuted into ^{128}I, with a half-life of 25 min through a (γ,n)-reaction using laser-generated bremsstrahlung.

9.3.4 Encapsulation of Radioactive Samples

The Nuclear Fuels unit at ITU has a long-established history in the design of encapsulation techniques for radioactive samples (Fig. 9.8). Encapsulation allows one to handle, store, transport, and perform experiments with radioactive samples in a safe and flexible manner and to avoid any contamination. One of the first techniques developed was the aluminum encapsulation of neptunium samples for Mössbauer spectroscopy studies. Since 1996, the technique has been extended to Ag encapsulation of radium-226 for applications in cancer therapy (alpha-immunotherapy). Up to 40 capsules have been produced with radium-226 in quantities ranging from 6 μg to 30 mg (from 2.2×10^5 to 1.1×10^9 Bq). Special containers in Al have been designed with the help of the Basic Actinide Research unit to contain beta-radiation from radioactive beta-sources, such as iodine-129 and technetium-99. In the case of the iodine, plexiglas encapsulation was used as an additional radiation barrier.

Radium chloride (^{226}RaCl$_2$) samples encapsulated in silver undergo proton irradiation in a cyclotron. The ^{225}Ac required for medical application is produced through (p,2n) with the radium. After irradiation, the target is dissolved and then treated in a conventional way in order to separate Ac from Ra, for example in ion exchangers. For encapsulation, the following facilities are used:

- DC tungsten inert gas welding equipment, using pulsed current to weld Ag capsules of thickness 0.25 mm.
- AC/DC tungsten inert gas welding equipment for welding of aluminum (purity 99.99%).
- Dedicated glovebox chamber attachment, which is disposable, allows the welding equipment to be used with various gloveboxes without being exposed to contamination.
- Specially developed handling devices that allow fast, precise filling and mounting of the capsules for welding, thereby minimizing the exposure to high dose rates, especially with radium-226.
- Quality control by radiography and helium leak test.
- Production facilities for capsule components of diverse materials and geometries. These include cold pressing of high-purity aluminum components, which are difficult to produce by normal machining processes.

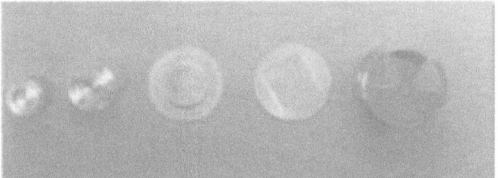

Several kind of capsules designed and built at ITU. Starting from the left, the second one is the actual aluminum capsule for Tc. The third and the fourth are made of silver and used for ^{226}Ra enncapsulation and recently for ^{238}U and ^{232}Th for foreseen proton-induced fission laser experiments.

Tungsten Inert Gas Weldilng Equipment, mounted on a glovebox chamber attachment

Fig. 9.8. Encapsulation of radioactive samples at the Institute for Transuranium Elements (ITU), Karlsruhe

9.3.5 Laser-Induced Heavy Ion Fusion

In a recent series of experiments with the VULCAN laser, at intensities of 10^{19} W/cm^2, beams of energetic ions were produced by firing the laser onto a thin foil primary target (see Fig. 9.5). The resulting ion beam then interacts with a secondary target. If the ions have enough kinetic energy, it is possible to produce fusion of the ions in the beam with atoms in the secondary target. Heavy ion beams were generated from primary targets of aluminum and carbon. Secondary target material consisted of aluminum, titanium, iron, and zinc niobium and silver. The heavy ion "blow-off" fused with the atoms in the secondary target creating compound nuclei in highly excited states. The compound nuclei then deexcited to create fusion products in the secondary target foils. These foils were then examined in a high-efficiency germanium detector

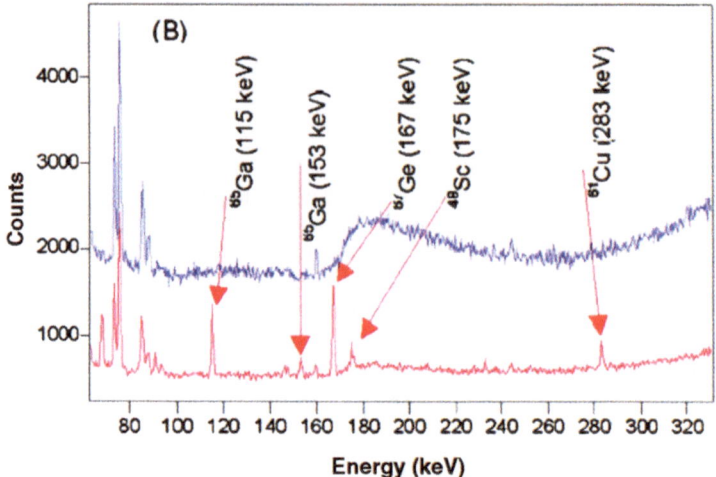

Fig. 9.9. The main reaction products identified by their characteristic gamma emission for a Ti plate exposed to Al blow-off. Blue spectrum: "cold" target, red spectrum, heated target (391°C). Fusion products are much more evident in the heated target

to measure the characteristic gamma radiation produced by the radioactive decay of short-lived fusion product nuclides. Typical spectra are shown in Fig. 9.9. Figure 9.9 also shows the results of experiments involving cold and heated targets. The target here was aluminum, and the secondary titanium. The spectrum in blue is that taken for the aluminum target at room temperature, and the red spectrum is that of an aluminum target heated to 391°C. For the heated target, many more fusion products are evident which are not observed in the cold target. This is attributed to the heating of the target to remove hydrocarbon impurities. When these layers are removed, heavier ions are accelerated more readily and to higher energies.

9.3.6 Laser-Generated Protons and Neutrons

Recently, (p,xn) reactions on lead with the use of very high intensity laser radiation has been demonstrated [21, 22]. Laser radiation is focussed onto a thin foil to an intensity of 10^{20} W/cm^2 to produce a beam of high-energy protons. These protons interact with a lead target to produce (p,xn) reactions. The (p,xn) process is clearly visible through the production of a variety of bismuth isotopes with natural lead. Such experiments may provide useful basic nuclear data for transmutation in the energy range 20–250 MeV without recourse to large accelerator facilities.

At low energies ($\leq$50 MeV), the de Broglie wavelength of the proton is larger than the size of individual nucleons. The proton then interacts with the entire nucleus and a compound nucleus is formed. At high proton energies ($\geq$50 MeV), the de Broglie wavelength is of the order of the nucleon dimensions. The proton can interact with single or a few nucleons and results in direct reactions. These latter reactions are referred to as spallation nuclear reactions and refer to nonelastic interactions induced by a high-energy particle in which mainly light charged particles and neutrons are "spalled," or knocked out of the nucleus directly, followed by the evaporation of low-energy particles as the excited nucleus heats up. Current measurements on the feasibility of proton-induced spallation of lead and similar materials focus around the need to measure nuclear reaction cross sections relevant to accelerator-driven systems desirable for use in the transmutation of long-lived radioactive products in nuclear waste. The neutron production from the spallation reaction is important for defining the proton beam energy and target requirements. However, the measurements being undertaken require high-power accelerators to generate the proton beam. In the present work, the proton beam is generated by a high-intensity laser rather than by an accelerator.

The recently developed petawatt arm of the VULCAN Nd:glass laser at the Rutherford Appleton Laboratory, U.K., was used in this experiment. P-polarized laser pulses with energy up to 400 J, wavelength approximately 1 m, and average duration 0.7 ps, were focussed onto foil targets at an angle of 45° and to an intensity of the order of 4×10^{20} W/cm^2. A typical spectrum resulting from the proton activation of lead to produce bismuth isotopes is shown in Fig. 9.10 (the relevant section of the nuclide chart is shown in Fig. 9.11).

The protons originate from H_2O and hydrocarbon contamination layers on the surface of solid targets. Secondary catcher activation samples were positioned at the front of the target (the "blow-off" direction). Energetic protons

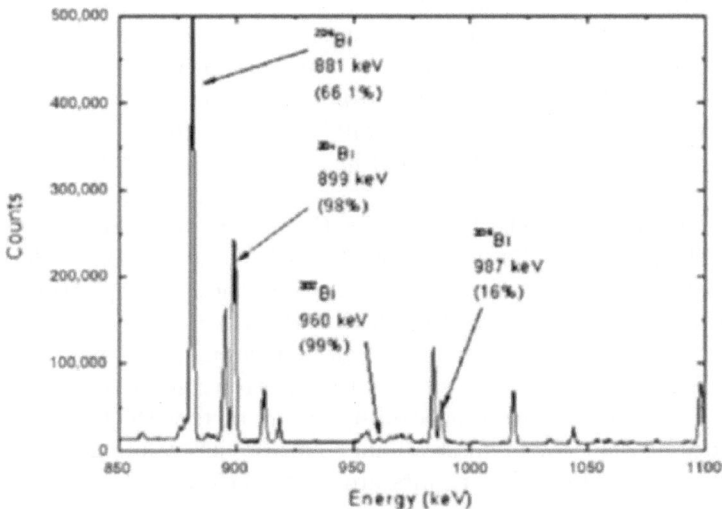

Fig. 9.10. Preliminary identification of bismuth isotopes produced through (p, xn) reactions in lead

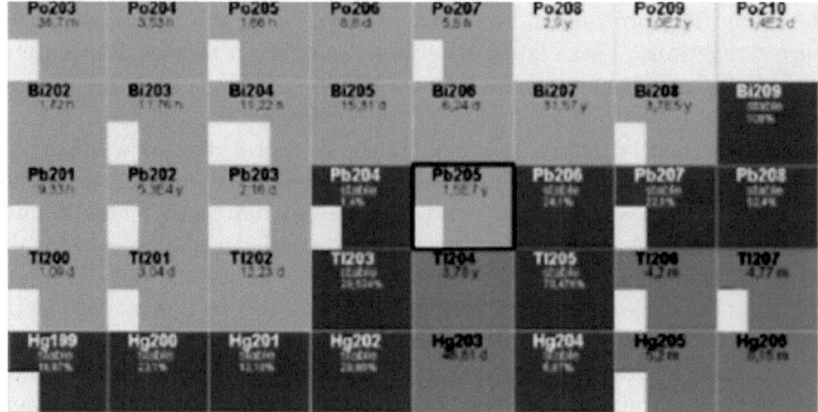

Fig. 9.11. Nuclide chart [26] showing the location of lead and bismuth isotopes

accelerated from the primary target foil can induce nuclear reactions in these activation samples. From the proton-induced reactions on lead, the isotopes $^{202-206}$Bi were identified using the main emission lines.

9.3.7 Laser Activation of Microspheres

Nano-encapsulation of chemical agents is well known in the pharmaceutical field. Nano-radiotherapy is a technique in which nanoparticles can be made radioactive and then used in cancer therapy [23, 24]. Nanospheres are relatively easy to manufacture and the isotope to be activated is chosen depending on the type and size of the tumor. The particles are activated by neutron irradiation in a nuclear reactor. Typically, durable ceramic microspheres containing a large amount of yttrium and/or phosphorus are useful for in situ radiotherapy of cancer, since the stable ^{89}Y and/or ^{31}P in the microspheres can be activated to emitters ^{89}Y with a half-life: 2.7 d and/or ^{32}P with a half-life of 14.3 d.

Recently, microparticles have been activated in a ultra-high-intensity laser field for the first time [25]. Microparticles of ZrO_2 and HfO_2, with diameters of approximately 80 μm, were irradiated using the high repetition rate laser at the University of Jena (see Figs. 9.12 and 9.13).

Focussing the laser beam in the gas jet results in a high-temperature plasma. Relativistic self-focussing of the electrons gives rise to a directed, pulsed, high-energetic electron beam which interacts with the primary target to produce high-energy bremsstrahlung. This bremsstrahlung is then used for the particle activation. The results are shown below for both the zirconium and hafnium microparticles.

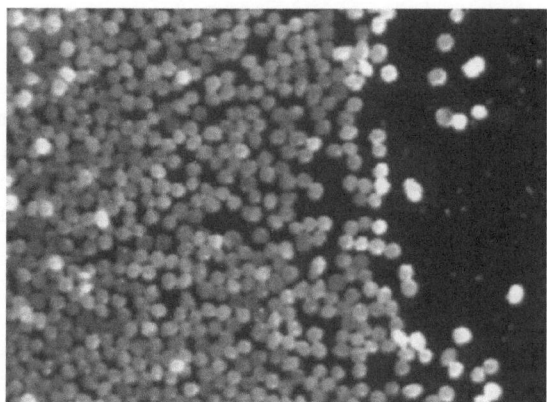

Fig. 9.12. Zirconium oxide microparticles. The particle diameters are in the range 95–110 μm

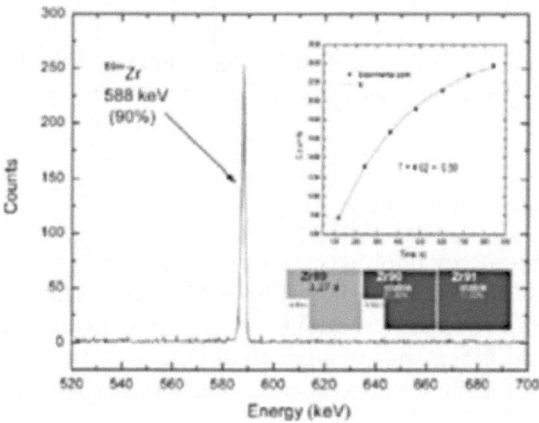

Fig. 9.13. Spectra taken 10 min after the Zr irradiation showing the main line of ^{89m}Zr from the ^{90}Zr$(\gamma,$n$)^{89m}$Zr reaction. The inset shows the position of the isotopes in the nuclide chart (from Nuclides.net [26])

9.3.8 Tabletop Lasers for "Homeland Security" Applications

Recently there has been a renewal of interest in photo-nuclear processes motivated by a number of applications. These applications include electron accelerators, shielding studies, radioactive nuclear beam production, transmutation of nuclear waste, nondestructive characterization of waste barrels, detection of nuclear material via photo fission, etc. A collaboration of the CEA and LLNL has been started to construct a photo-nuclear data library for the CINDER'90 calculation code, for example. The IAEA and NEA are also pursuing this actively. Also proton-induced reactions are of great practical interest (see below).

Detection of Explosives

Protons with energy of 1.75 MeV can be used to generate monoenergetic 9.17 MeV photons through the reaction $^{13}C(p,\gamma)$ ^{14}N. The compound nucleus $(^{14}N)^*$ produced is in an excited state and decays by the emission of characteristic prompt gamma radiation. These resonance photons can then be used to detect explosives (which also contain ^{14}N) by transmission or scattering of these resonance 9.17 MeV photons (Fig. 9.14).

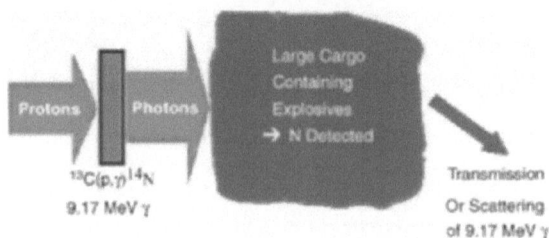

Fig. 9.14. Gamma resonance technology for explosive detection. (P. Oblozinsky, CSEWG-USNDP Meeting: Nuclear Data for Homeland Security [27])

Detection of Fissile Material

Neutrons or high-energy photons can be used to actively interrogate containers to detect special nuclear materials (Fig. 9.15). The technique is based on the use of high-energy gammas (>3 MeV) from the beta decay of fission fragments. The gamma yield is ten times higher than that of beta-delayed neutrons and gammas escape from hydrogenous cargo much easier than neutrons.

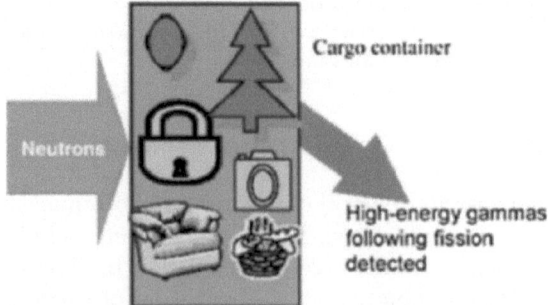

Fig. 9.15. Active (neutron and gamma) interrogation of fissile material. (E. Norman, CSEWG-USNDP Meeting: Nuclear Data for Homeland Security [27])

Nuclear Materials Detection

This is a technique for detection of nuclear materials (U, Pu, Be, D, ^{6}Li) with low-energy protons. Low-energy (<5 MeV) protons are used to produce 6–7 MeV photons through the reaction ^{19}F(p,$\alpha\gamma$)^{16}O. These monoenergetic gammas are above the threshold for photo-fission and photo-nuclear reactions.

9.4 Conclusions

The future development of the field of laser transmutation will benefit from the currently fast evolution of high-intensity laser technology. Within a few years, compact and efficient laser systems will emerge, capable of producing intensities exceeding 10^{22} W/cm^2 with repetition rates of 1 shot per minute and higher. These laser pulses will generate electron and photon temperatures in the range of the giant dipole resonances and open the possibility of obtaining nuclear data in this region. These laser experiments may offer a new approach to studying material behavior under neutral and charged particle irradiation without resource to nuclear reactors or particle accelerators.

References

1. M. Howarth: *Poiner Research on the Atom.* (London, 1958), pp. 3–84
2. A. Fleck: "Frederic Soddy" in Biographical Memoirs of Fellows of the Royal Society **3**, 203–216 (1957)
3. E. Rutherford, J.E. Petavel: Br. Assoc. Advan. Sci. Rep. A 456 (1906)
4. E. Segre: Phys. Rev. **71**, 274 (1947)
5. G.T. Emery: Ann. Rev. Nucl. Sci. **22**, 165 (1972)
6. H. Mazaki: J. Phys. E, Sci. Instrum. **11**, 739–741 (1978)
7. K. Ader, G. Bauer, V. Raff: Helv. Phys. Acta **44**, 514 (1971)
8. W. Rubinson, M.L. Perlman: Phys. Lett. B **40**, 352 (1972)
9. K.W.D. Ledingham et al.: Science **300**, 1107–1111 (2003)
10. K.W.D. Ledingham et al.: Phys. Rev. Lett. **84**, 899 (2000)
11. H. Schwoerer et al.: Europhys. Lett. **61**, 47 (2003)
12. M.J. Feldman, R.Y. Chiao: Phys. Rev A **4**, 352–358 (1971)
13. T. Tajima, G. Mourou: Phys. Rev. Spec. Topics – Accelerators Beams **5**, 031301-1 (2002)
14. O.E. Martinez et al.: J. Opt. Soc. Am. A**1** 1003–1006 (1984)
15. P. Main et al.: IEEE J. Quant. Electr. **24**, 398–403 (1988)
16. C.K. Rhodes: Science **229**, 1345–1351 (1985)
17. K. Boyer, T.S. Luk, C.K. Rhodes: Phys. Rev. Lett. **60**, 557–560 (1988)
18. J. Magill et al.: Appl. Phys. **B**, 1–4 (2003)
19. K.W.D. Ledingham et al.: J. Phys. D: Appl. Phys. **36**, L79–L82 (2003)
20. F. Ewald et al.: Plasma Phys. Control. Fusion **45**, 1–9 (2003)
21. P. McKenna et al.: Phys. Rev. Lett. **94**, 084801 (2005)
22. T. Žagar et al.: New J. Phys. **7**, 253 (2005)

23. See http://www.ualberta.ca/∼csps/JPPS3(2)/M.Kumar/particles.htm
24. See http://www.nea.fr/html/pt/docs/iem/madrid00/Proceedings/Paper41.pdf
25. J. Magill et al.: Joint Research Centre Technical Report JRC-ITU-TN-2003/08 (March 2003)
26. J. Magill: *Nuclides.net: An Integrated Environment for Computations on Radionuclides and their Radiation.* (Springer, Heidelberg 2003), http://www.nuclides.net.
27. Proceedings of the CSEWG-USNDP 2004 Meetings, November 2–5, 2004; http://www.nndc.bnl.gov/proceedings/2004csewgusndp/

10

High-brightness γ-Ray Generation for Nuclear Transmutation

K. Imasaki[1], D. Li[1], S. Miyamoto[2], S. Amano[2], and T. Mochizuki[2]

[1] Institute for Laser Technology, 2-6, Yamada-Oka Suita, Osaka, 565-0871 Japan
[2] Laboratory of Advanced Science and Technology for Industry, University of Hyogo, Ako, Hyogo Japan
kzoimsk@ile.osaka-u.ac.jp

Abstract. In this article, a generation of high-brightness γ-ray sources and application of these sources to transmutation of nuclear waste are discussed. Recent developments in laser and optical technology allow us to store photons in a cavity. Interaction of an electron beam with stored photons results in an enhancement of the γ ray generation. Such γ-ray sources are expected to have high efficiency. We have studied a conceptual design of an integral system for nuclear waste transmutation of both long-lived fission products (FP) and transuranic (TRU) waste and discuss the energy balance in this system. We performed a small-scale experiment with a low-energy electron beam and stored photons in a supercavity. These results were in good correspondence with the predictions based on the cavity storage rate and electron beam energy. Experiments for preliminary nuclear transmutation are under way on 1.5 GeV new SUBARU electron storage ring at the University of Hyogo.

10.1 Introduction

Transmutation of long-lived fission products and transuranic (TRU) waste is an option to reduce the burden of nuclear waste in geological repositories. There have been several proposals for such transmutation scenarios. Using γ-rays, the transmutation through excitation of the giant nuclear resonance had been proposed. This method uses the bremsstrahlung γ-rays generated in a target by the interaction of a high-energy electron beam. The energy conversion efficiency, however, from the electron beam to γ-ray is not high and the spectrum is wide. This results in a poor coupling of energy to the giant resonance for the transmutation. Here we proposed a new approach to γ-ray nuclear transmutation [1].

Recent developments in laser and optical technology make photon storage in a cavity a possibility. Generation of high-brightness radiation due to the accumulated photons interacting with high-energy electrons by Compton scattering is expected, and such radiation can be tuned for applications in target nuclei. In addition to these advantages, the total conversion efficiency

K. Imasaki et al.: *High-brightness γ-Ray Generation for Nuclear Transmutation*, Lect. Notes Phys. **694**, 147–167 (2006)
www.springerlink.com   © Springer-Verlag Berlin Heidelberg and European Communities 2006

from electric power to the γ-ray can be high enough to get the energy balance for nuclear transmutation. The electron accelerator is compact and has high efficiency. These facts make a good cost performance for the transmutation [2, 3].

In Sect. 10.2, the principles of this scheme are presented. Preliminary experiments of photon storage and interaction of such photons with an electron beam were performed. Interaction of γ-rays with the target is discussed. A nuclear transmutation experiment has been performed in the New SUBARU storage ring generating 17 MeV laser Compton γ-rays. This is described in Sect. 10.3. In Sect. 10.4, we consider the application to transmutation. The conceptual system design of the generation of high-brightness γ-rays with high efficiency and transmutation with energy recovery is discussed. A summary is presented in Sect. 10.5.

10.2 Principles of this Scheme

The principles of this scheme are shown in Fig. 10.1. Storage of photons is performed by a stable supercavity composed of high reflectivity and extremely low loss mirror pair. A high-energy and high-brightness electron beam with high average current is supplied by an electron storage ring with bypass orbit. The Compton scattering with high efficiency in the so-called supercavity generates high-brightness γ-rays. The γ-ray energy is tuned by the energy of the electron beam and the laser photons and coincides with the peak of E1 giant resonance. Neutrons are generated in this process, which can also be used for transmutation.

10.2.1 Laser Photon Storage Cavity

An experiment for proof of principal photon storage and interaction with electrons was performed. Figure 10.2 shows picture of the supercavity and its structure for the experiment [4].

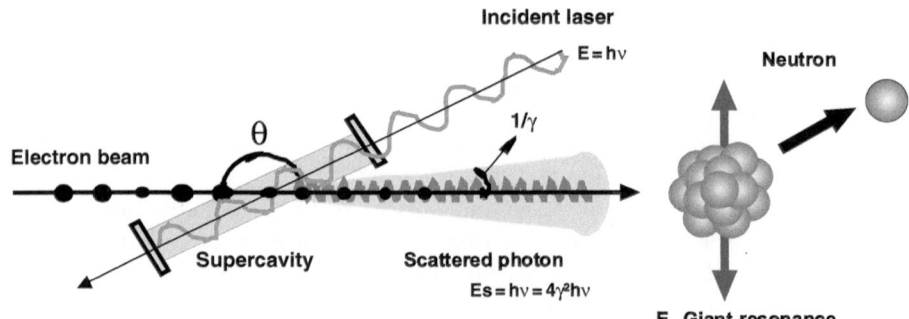

Fig. 10.1. Principles of this scheme

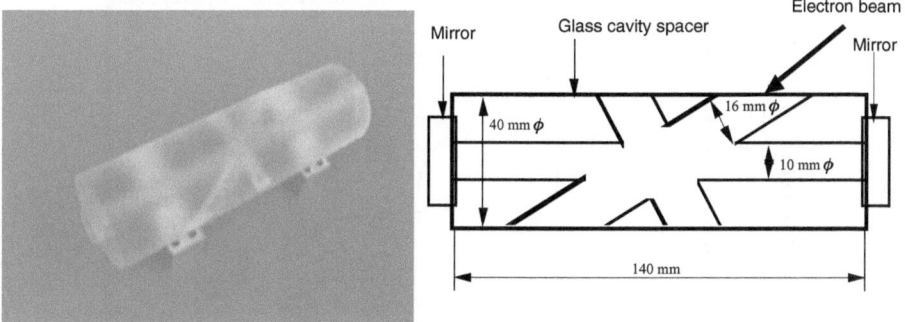

Fig. 10.2. Picture of the photon storage cavity and its structure

Laser photon storage cavity is a Fabry-Perot interferometer with a high-quality mirror pair. The mirrors are required to have not only a high reflectivity ($R = 99\% \sim 99.999\%$) but also a low loss ($1 \sim 10$ ppm). To evaluate the characteristics of the optical cavity, reflectivity R, transference T, and loss A are important parameters, and they relate to each other through $R + T + A = 1$. The cavity transmittance and reflectivity can be written as follows:

$$\eta_T = \left(\frac{T}{A+T}\right)^2 \quad \text{and} \quad \eta_R = \left(\frac{A}{A+T}\right)^2.$$

The estimated storage rate in the cavity is η_T/T shown in Fig. 10.3. We can expect a photon storage rate of up to 10^5, using high-quality mirror.

We used the cavities and performed the experiments for various mirrors with glass and metal cavity spacer to store the laser photons. The results are shown in Fig. 10.3 and compared with the theoretical curves. The storage rate was measured by a ring down method monitoring the decay rate of laser photons in the cavity. The $\star$ are the results obtained by several kinds of mirror pairs with glass and metal cavity supporter shown in Fig. 10.3 and points $\bigcirc$ are the results using a smaller cavity.

The field generated inside the cavity can be given by the ratio of transmitted power divided by the transmittance. We obtained a storage rate around 10000 on the cavity shown in Fig. 10.3 with a stability of more than 10 h. We also measured the scattered photon number from Compton scattering to evaluate the laser intensity in the cavity as shown in Fig. 10.4. Both experimental results of storage rate agreed very well with each other [5].

10.2.2 Photon–Electron Interaction

Interaction of the photons with the e-beam is induced by Compton scattering. This cross section is very small. So normal Compton scattering requires very high power laser with several tens of GJ of average energy for the purpose of transmutation. Such a laser is not realizable. But in the cavity with high

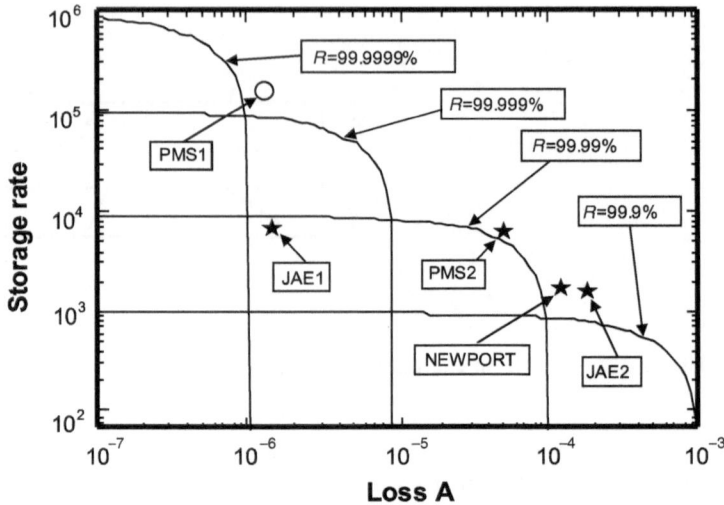

Fig. 10.3. Theoretical curves and experimental results of photon storage cavity. PMS, JAE, and so on are the names of mirror manufacturers

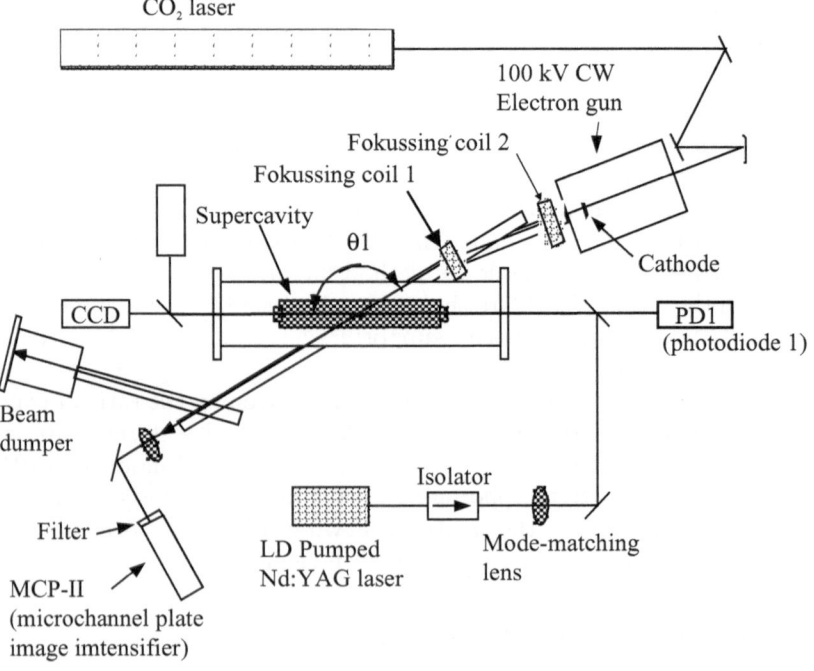

Fig. 10.4. Compton scattering experiments in a cavity. Electron energy was 100 keV and emitted by the laser-heated cathode. Supercavity to storage the laser photon and LD pumped Nd:YAG laser was used

storage rate noted above, it is possible to obtain such high laser intensity using conventional laser. So the generation of γ-rays for the transmutation can be considerable by the application of new laser technologies [2].

Figure 10.1 shows the schematic picture of Compton scattering in the supercavity. The γ-ray energy of the scattered photon becomes $4\gamma^2$ times that of the initial incident photon energy $h\nu$ with the solid angle of $1/\gamma$. The interaction angle θ is to be taken near 2π to get a maximum interaction. The direction of the scattered photon is given by the electron beam direction. The target of the transmutation will have a diameter smaller than 1 cm.

The laser intensity in the cavity is so high that Compton scattering becomes nonlinear or multiple scattering occurs. The power density threshold for the nonlinear Compton scattering for 1 μm laser is $10^{22}\,\mathrm{W/m^2}$ [3]. It is possible not to exceed this value to avoid the wide energy spread of the electron beam in the storage ring. But multiple scattering is induced in the cavity. Typical spectra of the electron beam calculated for various laser intensities after the scattering are shown in Fig. 10.5. The initial electron energy was 1 GeV for 1 μm laser. Clear effects of multiple scattering were observed for laser intensities more than $10^{18}\,\mathrm{W/m^2}$ [4].

For the storage ring, to keep the electron beam in the stable orbit is important for energy balance. So we have to use a bypass system to hold the beam for the multiple scattering stages. The electron beam, which interacts with photon beam in a single bypass, is switched into normal orbit of the storage

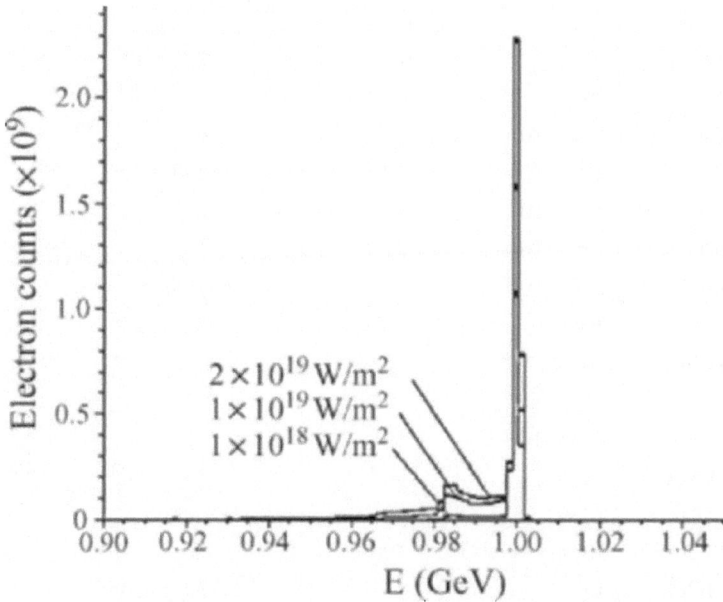

Fig. 10.5. Electron spectrum after Compton scatter in storage ring with the laser power density in the range $10^{18}\,\mathrm{W/m^2}$ to $2 \times 10^{19}\,\mathrm{W/m^2}$

ring and is cooled to circulate around several times without interaction. After this the electron beam interacts again in the bypass. The system repeats this to keep the beam in the ring. A detailed simulation is under way to design the electron storage ring for the practical transmutation system.

10.2.3 Target Interaction

Direct Target for γ-Ray

Nuclides candidates of the transmutation are listed in Table 10.1. These results typically from a 1-year operation of a 1 GWe reactor [7]. The γ-rays will induce several reactions in the target. Pair creation is dominant and is increases by Z^2. So lower Z target such as FPs are better for the giant resonance. Among them, iodine is a suitable target because it is transmuted into Xe in a few tens of minutes and can be separated easily from the others.

At the same time, neutrons are generated when the transmutation is occurs by gamma photons. These neutrons will have high density and induce a second reaction for the TRU and other FP placed around the direct target. Carbon

Table 10.1. Typical nuclear waste from the nuclear reactor (1 Gwe)

Nuclei	Half Decay (year)	Neutron Cross Section (b)	Production (Ci/year)	Amount (kg/year)
FP				
85Kr	11	1.7	3.0×10^5	0.79
90Sr	29	0.014	25×10^6	17.8
93Zr	1.5×10^6	2.6	61	24.0
99Tc	2.1×10^5	20	433	25.5
107Pd	6.5×10^6	1.8	3.6	7.0
129I	1.6×10^7	27	1.0	5.8
135Cs	2.3×10^6	8.7	13.5	11.7
137Cs	30	0.25	3.5×10^6	39.5
151Sm	90	15,000	1.1×10^4	0.4
TRU				
237Np	2.1×10^6	181	11	14.4
241Am	432	603	5.0×10^3	1.46
243Am	7380	79	601	3.03
243Cm	28,5	720	55	0.01
244Cm	18	15	5.8×10^4	0.72
245Cm	8500	2,347	4.1×10^3	0.03

is another possible candidate as a direct target and neutron source. Carbon is transmuted to stable boron after the γ-neutron reaction. Boron decays into alpha particles and protons for the next γ-neutron reaction. There is a possibility to get a larger enhanced reaction rate in carbon targets as we discuss below.

Reaction Rate

By the annihilation photons of positrons, cross sections of the E1 mode nuclear giant resonance for many nuclides have been measured precisely [6]. The scaling of total cross section and peak energy of reaction are studied below. We can use them to calculate the reaction rate for transmutation and design the transmutation system. The reaction rate R_{rea} of γ-rays for the transmutation is written as

$$R_{rea} = \langle \sigma_{gr}(E_p) \rangle / \langle \sigma_{pa}(E_p) + \sigma_{gr}(E_p) + \sigma_{co}(E_p) + \sigma_{pe}(E_p) \rangle , \qquad (10.1)$$

where $\sigma_{pa}(E_p)$ is the cross section for pair creation for gamma photons of energy E_F, $\sigma_{gr}(E_F)$ is a cross section of the giant resonance, $\sigma_{pe}(E_p)$ is the cross section for Compton scattering of electrons in the target and $\sigma_{co}(E_p)$ is the cross section for photoelectron production. A typical cross section of each process is shown in Fig. 10.6 for a FP target with typical γ-ray energy spectrum [8].

Curve a is a cross section for pair creation, curve b is that for Compton scattering by the electrons in the target, curve c is that for the giant resonance, and d is that for the photoelectron effect. Curve e is a typical gamma photon spectrum by E-beam and laser interaction in this case. Equation (10.1) is modified for the high Z material of FP or TRU as

$$R_{rea} = \langle \sigma_{gr}(E_p) \rangle / \langle \sigma_{pa}(E_p) \rangle . \qquad (10.2)$$

The normal reaction rate for FP is 3%. Better reaction rates are desirable to obtain an energy balance. For the high Z and medium Z targets, it is important to suppress pair creation to enhance the reaction.

On the other hand, low Z target as normal carbon, (10.1) can be approximately written as

$$R_{rea} = \langle \sigma_{gr}(E_p) \rangle / \langle \sigma_{gr}(E_p) + \sigma_{co}(E_p) \rangle . \qquad (10.3)$$

The suppression of Compton scattering by electrons in the target is dominant in this case. The enhancement was expected by appropriate magnetic fields in the target and polarized γ-rays for each case of pair creation and target electron Compton scattering. These are shown in Fig. 10.7. In the optimistic case, we can expect the reaction rate more than 5% for the carbon target by the suppression of Compton scattering of target electrons and pair creation of nuclei of both targets

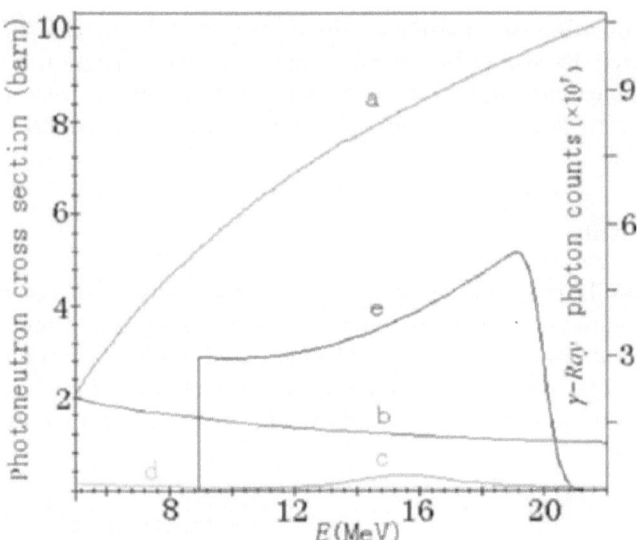

Fig. 10.6. Typical cross sections of processes in direct target interaction. Curve a is the cross section of pair creation, b is that for Compton scattering, c is that of giant resonance and d is the photoelectron effect. Curve e is a typical γ-ray spectrum by E-beam and laser interaction

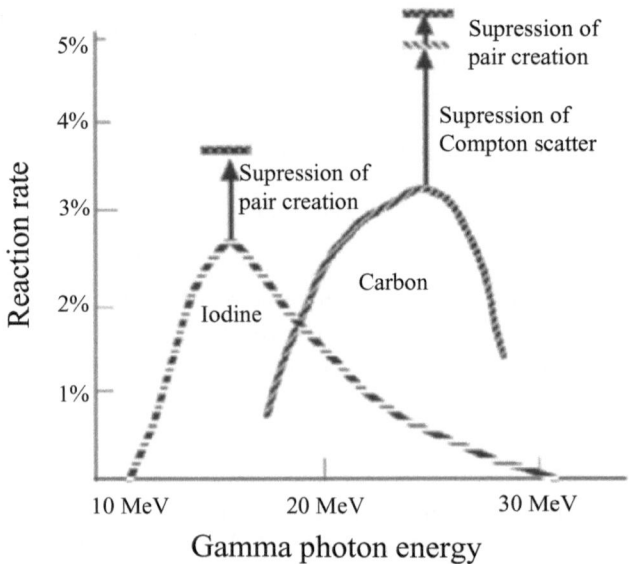

Fig. 10.7. Reaction rate of direct target for γ-ray and feasibility of enhanced interaction

10.3 Transmutation Experiment on New SUBARU

10.3.1 γ-Ray Generation for the Transmutation

The transmutation experiment is performed with 1.5 GeV New SUBARU storage ring. We can generate the γ-rays to induce the giant resonance and performed experiments in transmutation. The experimental configuration is shown in Fig. 10.8. BL 1 was the used beam line for γ-ray generation [9].

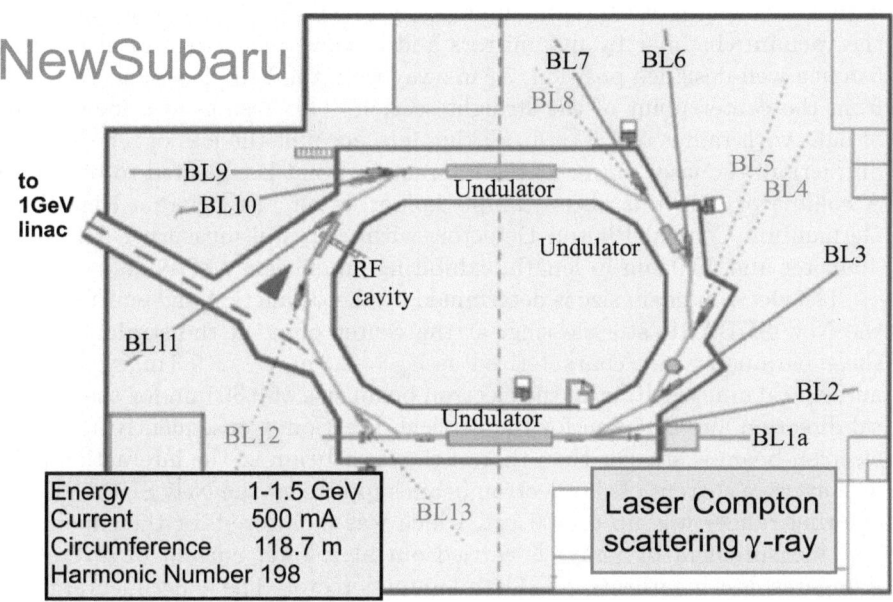

Fig. 10.8. New SUBARU Electron Storage Ring for transmutation experiments

We measure the γ-ray photon numbers and the energy spectrum by Ge detector to obtain the transmutation rate. One of the straight sections of the New SUBARU storage ring was chosen to realize the laser Compton scattering, where the electron beam collides with the incoming laser beam in a head-to-head manner. Thus the collisions between electrons and laser photons would give rise to higher energy photons, going along the incident electron moving direction in a forward cone of angle $1/\gamma, 0.5$ mrad in our experiment for the 1 GeV electron beam. A reflected mirror is located at the downstream end to guide the laser light travelling along the beam line through the interaction point designed at the center of the straight section, and the light is reflected out of the chamber by another upstream mirror. The produced γ-ray photons would go through the downstream mirror and reach the detector or irradiate the nuclear sample.

γ-ray photons are generated along the incident electron moving direction in a forward cone of angle $1/\gamma, 0.5\,\text{mrad}$ in our experiment for the $1\,\text{GeV}$ electron beam. A reflection mirror is located at the downstream end to guide the laser light travelling along the beam line through the interaction point designed at the center of the straight section, and the light is reflected out of the chamber by another upstream mirror. The produced γ-ray photons go through the downstream mirror and reach the detector or irradiate the nuclear sample.

The laser light is produced from a Nd:YAG laser operating at cw mode with a wavelength of $1.064\,\mu\text{m}$ and a power of $0.67\,\text{W}$. The light is guided into the vacuum chamber by five mirrors and a convex lens with focal length of $5\,\text{m}$ in a well-designed position, $7.5\,\text{m}$ away from the YAG laser and $15\,\text{m}$ away from the center point of the straight section. This results in a focused spot of light with radius of $0.82\,\text{mm}$. Taking into account the loss of reflection and diffraction, the laser power at the interaction point is expected to be $0.35\,\text{W}$. A collimator, which is also a sample holder, is set just before a High-Purity Germanium Coaxial Photon Detector, with a crystal measuring $64.3\,\text{mm}$ in diameter and $60.0\,\text{mm}$ in length, exhibiting an efficiency of 45%.

The electron beam size is determined by the β-function and emittance. For the New SUBARU storage ring, at the center point of the straight section, these parameters are characterized as $\beta_x = 2.3\,\text{m}$, $\beta_y = 9.3\,\text{m}$, $\varepsilon_x = 40\,\text{nm}$, and $\varepsilon_y = 4\,\text{nm}$, resulting in the electron beam size of $0.30\,\text{mm}$ for the horizontal direction and $0.19\,\text{mm}$ for the vertical direction. Consequently, the size of electron beam is smaller than that of the laser beam at the interaction point. The average current of the electron beam supplied by the New SUBARU storage ring ranges was up to $200\,\text{mA}$, which was monitored for the experiment. The measurement of γ-rays is carried out at a lower current of several milliamperes, lest it saturates the Germanium detector. The experimental results for Laser on and Laser off, detected by a collimator of $6\,\text{mm}$ in diameter are shown in Fig. 10.9. The apparent separation between the two signals of Laser Compton scattering γ-ray and the background presents a good signal-to-noise ratio. The maximum energy appears around $17\,\text{MeV}$, which is in agreement with the theoretical prediction.

We simulated the whole process of generated γ-ray photons passing through the reflection mirror, output window, collimator, and being detected by the Germanium detector, by employing the EGS4 code [10]. The EGS4 code is well known and widely used in the field of interaction of particles and material, taking into account many physical processes such as Bremsstrahlung production, pair production, Compton scattering, and photoelectric effect. The simulation curve is consistent with the experimental data as shown in Fig. 10.9. After processing the experimental data, we achieved the actual γ-ray photon luminosity of 2.5×10^5 counts/A/W/second. In conclusion, the γ-ray photon yield of 1.75×10^4 counts/second can be accomplished by our facility under the running condition of $I_e = 0.2\,\text{A}$ and $P = 0.35\,\text{W}$.

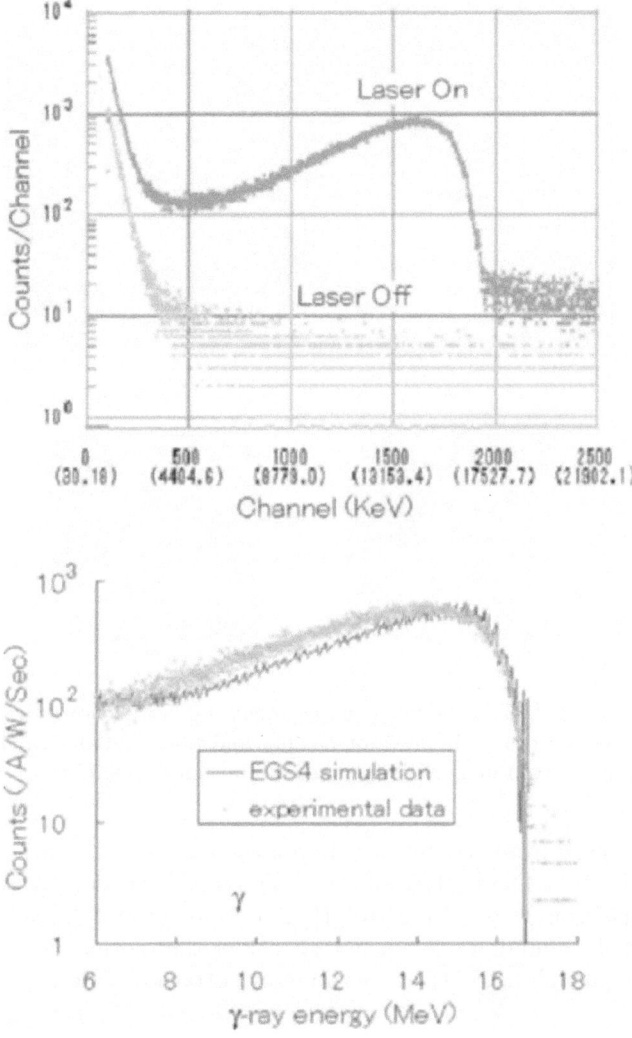

Fig. 10.9. γ-ray spectrum by Laser Compton scatter and comparison with calculated one

Polarized γ-rays, which are required to obtain an efficient interaction in the transmutation target, were produced by the polarized laser. The theoretical analysis predicts that the spatial distribution of intensity of laser Compton scattering γ-ray is connected with the polarization of initial laser photons. Circularly polarized or unpolarized initial photons give rise to an azimuthally symmetrical pattern in transverse distribution of intensity, whereas linearly polarized initial photon results in azimuthally modulation. In our experiment, the incoming laser photons were of linear polarization, and an image plate was

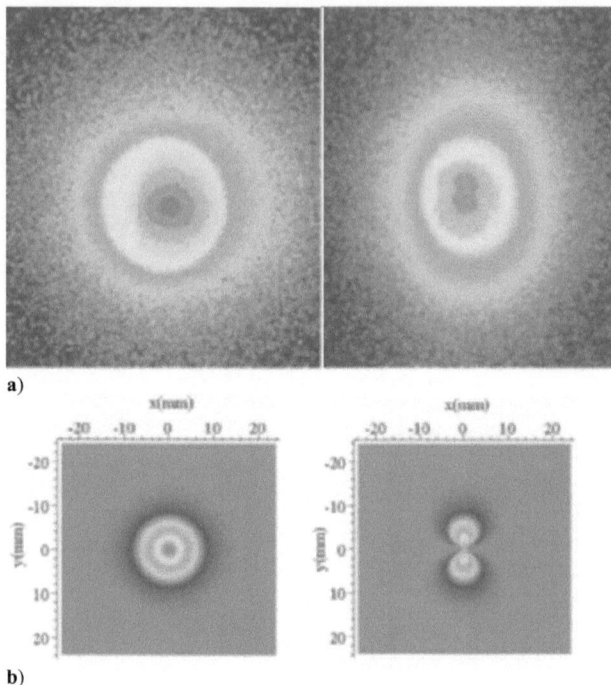

Fig. 10.10. The pattern of the γ-ray (experiment and calculation). (**a**) Experimental results of γ-ray pattern of 18 MeV for normal and linear polarized. (**b**) Calculated results of γ-ray pattern of 18 MeV for normal and linear polarized

placed 15 m away from the interaction point to detect the spatial distribution of produced γ-ray. The pattern of the γ-ray by the experiment and the calculation are shown in Figs. 10.10 a, b [11]. The experimental result corresponded well to that of calculated one when we consider that actual electron beam has a divergence, which made the experimental pattern obscure.

10.3.2 Nuclear Transmutation Rate Measurement

We used gold for the target. The transmutation process of the Au target is shown below and in Fig. 10.11.

Fig. 10.11. Experimental results of γ-ray pattern of 18 MeV for normal and linear polarized

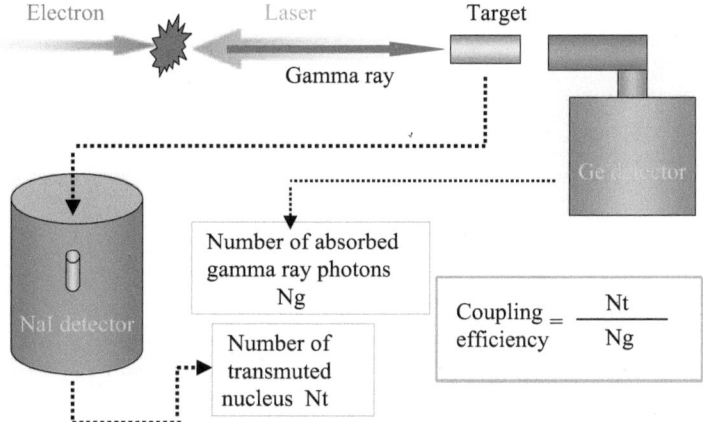

Fig. 10.12. Au target transmutation by 20 MeV γ-ray

The results of the experiment for counting the number of decays into Pt were performed. The energy of the radiation in the process of the decay of 196Au into 196Pt agreed with the theoretical prediction. The scheme of this experiment is shown in Fig. 10.12.

We performed investigations to explore the coupling efficiency of γ-rays to the nuclear giant resonance, which is defined as the transmutation rate per γ-ray photon. The coupling efficiency was derived by improving the one described in [4, 6] by considering geometrical structure of a cylindrical target as

$$\eta = \frac{N_0 \int_0^b \int_0^a \sigma_L(E)\sigma_g(E)e^{-\mu z} \cdot 2\pi dr}{\int \sigma_L(E) \cdot 2\pi dr}, \qquad (10.4)$$

where N_0 is the number of atoms per volume, $\sigma_L(E)$ is the cross section of laser Compton scattering defined by Klein–Nishina formula, $\sigma_g(E)$ is the cross section for nuclear giant resonance, μ is the total linear attenuation coefficient including the effects of photoelectron, Compton, and pair production as expressed in [8], a and b represent the radius and length of the cylindrical target, respectively, and E indicates the γ-ray energy.

Gold rods of 5 cm in length with a radius of 0.25 cm and 0.5 cm were adopted as the nuclear target in the present experiment and irradiated for a duration of 8 h on axis, 15 m away from the interaction point. The transmutation process for this target is shown in Fig. 10.12, and the main decay occurs from Au-196 to Pt-196, giving rise to radioactivity in the form of γ-ray photons with a peak energy of 355.73 keV in the energy spectrum. This radioactivity was measured by a NaI (TI)) detector, and the activity line was obtained and indicated a good agreement with the acknowledged half-life of 6.183 D. Through data processing, we concluded that the number of transmuted nuclei was 3.165×10^6 at the moment the irradiation was complete. On the other hand, by the Germanium detector, the absorbed Laser Compton

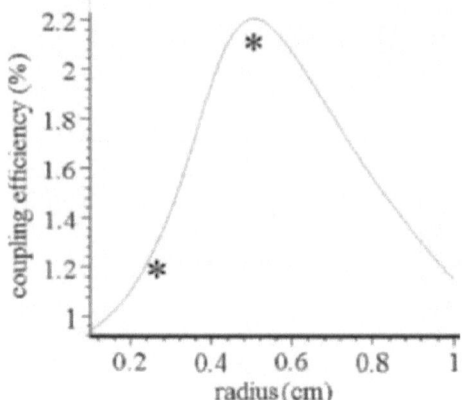

Fig. 10.13. Coupling of γ-ray to Au targets

γ-ray photons by the target during the irradiation was determined as 2.95×10^8. Hence, the coupling efficiency of γ-ray to nuclear giant resonance was derived as 1.1% and 2%. Actually, this value should be lower than the real value because the attenuation of γ-ray from the radioactivity inside the target was not involved, and future experiment would provide a more accurate estimation. However, the experimental result is close to the theoretical analysis as shown in Fig. 10.13 [12]. Now we are measuring the neutron spectrum from this reaction to understand the energy balancing on this method.

10.4 Transmutation System

A model of the system to generate high-brightness γ-ray for the transmutation is shown in Fig. 10.14. The most important point of the system is the efficiency and the low cost for an economical transmutation of nuclear waste. Here, the energy flow is discussed. The most important parameter is the γ-ray generation efficiency η_g.

10.4.1 γ-Ray Generation Efficiency

In a model as shown in Fig. 10.14, the γ-ray generation efficiency η_g can be written as

$$\eta_g = P_g[P_0 + P_b\tau_i/\tau_L + (nP_{sr} + P_g)/\eta_a + P_L/(\eta_L M)]^{-1} . \tag{10.5}$$

Here, P_g is γ-ray power (= electron energy loss), P_0 is power required for operation of storage ring including a utility and so on, P_b is electron beam power during the injection into the ring, τ_i is injection duration of the electron beam into storage ring, τ_L is Beam life time in the storage ring, P_{sr} is

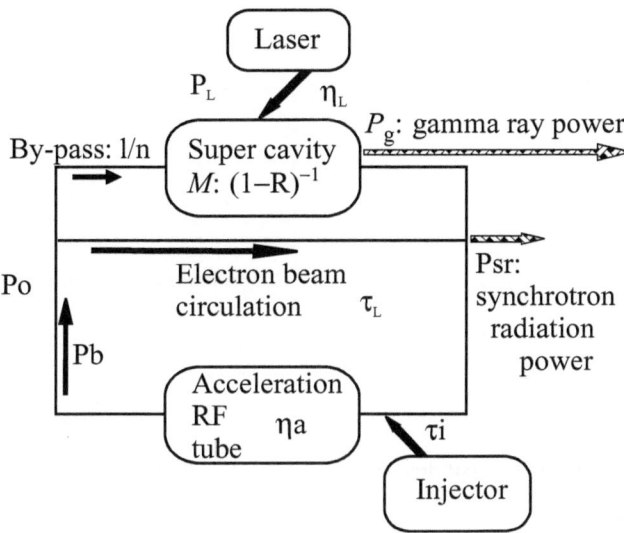

Fig. 10.14. Model of transmutation system by high-brightness γ-ray with high efficiency

synchrotron radiation power, n is a number of circulation times per one interaction, η_a is acceleration total efficiency including a klystron and so on, P_L is the interaction laser power in the supercavity, η_L is efficiency of laser for injection to the cavity, M is accumulating rate in the cavity $= (1 - R)^{-1}$, and R is a reflectivity of the mirror with extremely low loss. This equation can be rewritten approximately as

$$\eta_g = \eta_a[(nP_{sr}/P_g) + 1 + P_L/(P_g\eta_L M)]^{-1}(= \eta_a). \qquad (10.6)$$

As for typical parameters for superconduction accelerator, efficiency is very high as $\eta_a = 0.8$, which strongly depends on the efficiency of the accelerating tube. When we use a superconduction tube, we can expect very high efficiency.

10.4.2 Neutron Effect

The energy levels of the target nuclei are shown in Fig. 10.15. The giant resonance is induced when the energy of the γ-ray exceeds E_{thr} and emits a neutron. The neutron spectrum can be estimated as follows when we irradiate γ-ray around 20 MeV to the typical FP target. This analytic result corresponded well with that of the simulation.

The neutrons can cause the succeeding transmutations as shown in Fig. 10.16. The second target is composed of TRU and neutron multiplier. The reaction in the second target multiplies the neutron number and leads to energy balance in the system. The second target is a kind of subcritical fissionable blanket composed of mainly TRU. Reactions are induced three or

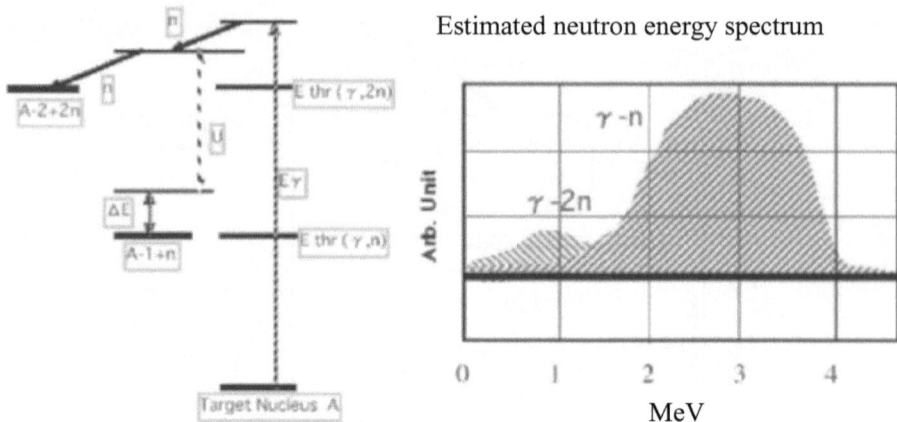

Fig. 10.15. Typical nuclear levels for γ-rays and estimated neutron spectrum for the processes

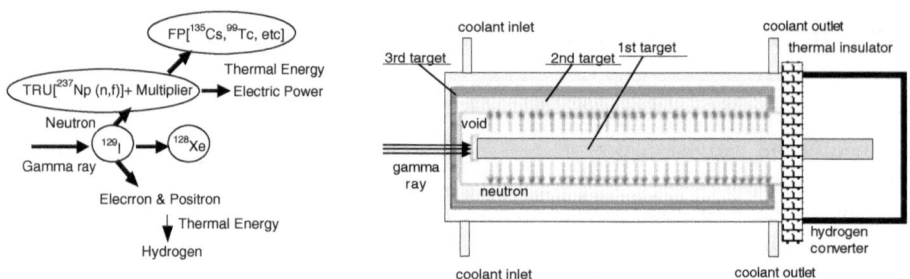

Fig. 10.16. Processes for targets and the rough drawing of the structure of targets

four times by MeV neutrons and lead to heat energy, which results in energy balance of the system.

The neutron number is so large that we can expect to obtain a high transmutation rate. In addition, the outer of the second target, the third target of FP as Tc, Cs, and so on, can be set for the neutron absorber to absorb the neutrons. Figure 10.17 shows an estimated neutron density in this case.

The first target is heated by electrons and positrons generated by the pair creation. This heat energy density is high enough to make hydrogen efficiently.

10.4.3 System Parameters

From the discussions, we can estimate the parameters of the actual system to transmute the long life FP and TRU as shown in Tables 10.2 and 10.3.

The parameters are not far from present-day technology future. The photon storage cavity is used for this estimation with a length of 1 m and the reflectivity of 4N mirror shown in Fig. 10.18. This system can transmute the long-lived FPs and TRUs from the 5 reactors of 1 GWe output.

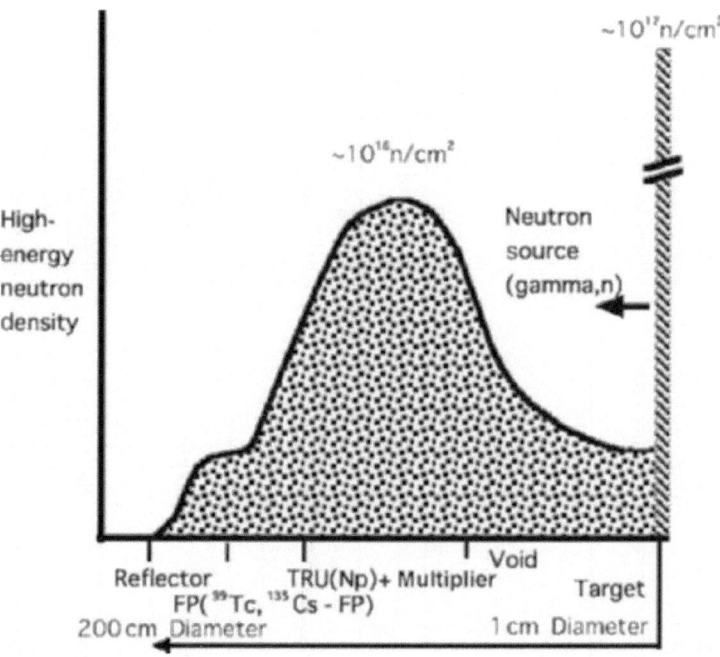

Fig. 10.17. Neutron density of each stage of the target structure. Target is 1 cm in diameter and 100 cm in length. There is a void between the center and second target, composed of TRU and neutron multiplier. The third target, composed of FP, absorbs neutrons from the second target

Table 10.2. Parameters of transmutation system

Component	Parameters	Requirements	State	Notes
Electron storage	Energy	3 GeV	–	Bypass
Ring with bypass	Current	15 A/beam AV.	2 A	
4 beams	Frequency	800 mHz	–	Energy
x5 machine	Acceptance	3%	2%	Circulation
CW CO_2 laser	Power	500 kW	200 kW	
x5 machine				
Photon	Accumulation rate	8000	7000–	CO_2 laser
Accumulation	Path number/unit	20	8000	multi-path
Cavity	Unit number/beam	10	for YAG laser	

Table 10.3. Transmutation system with various cases

Transmutation System			Case 1 Normal Interaction	Case 2 Energy Recover by TRU Blanket	Case 3 Case 2+hydrogen Production	Case 4 Carbon Target Enhanced Interaction
γ-Ray photon		Energy	17 MeV	17 MeV	17 MeV	17 MeV
		number/s	2×10^{21}	2×10^{21}	2×10^{21}	10^{21}
		reaction rate required	3%	3%	3%	6%
		total energy	6 GW	6 GW	6 GW	2 GW
Target (50 reactors)	direct	129Iodine	290 kg	290 kg	290 kg	carbon target 100 kg
	2nd sub-Critical	TRU (Np, Am.Cm) 2 MeV neutron	600 kg	1000 kg	1000 kg	TRU 1500 kg
	3rd	FP-Tc	None	2000 kg	2000 kg	Tc-Iodine 2000 kg Cs 2000 kg
Initial cost (estimated)			1000 m$	2000 m$	2500 m$	2000 m$
Operation cost /year (estimated)			+500 m$ (electric power cost)	0 (energy balance)	−200 m$ (gain) (hydrogen output & target fabrication)	−200 m$ (hydrogen output & target fabrication)

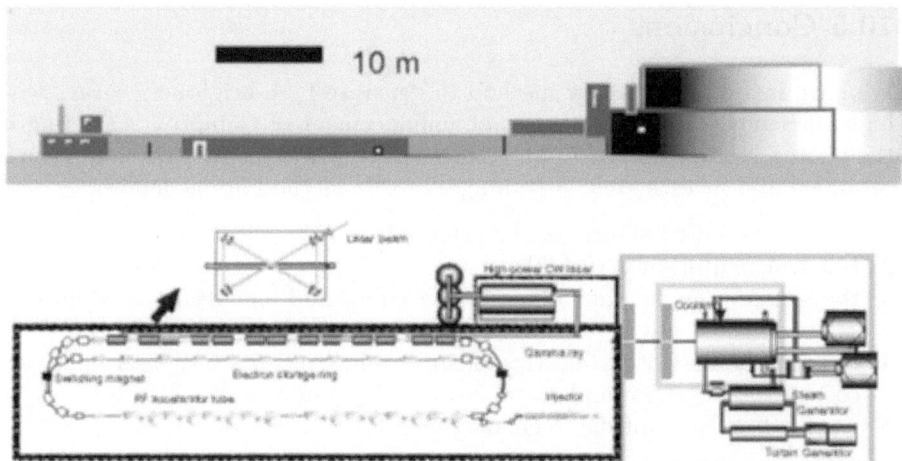

Fig. 10.18. Schematic picture of the transmutation facility

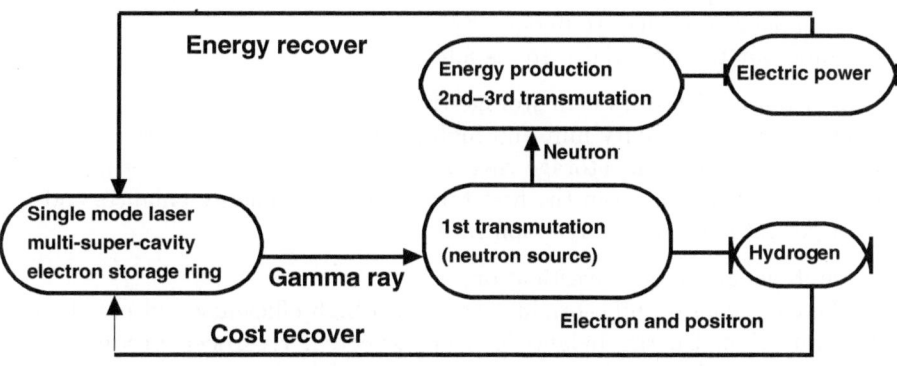

Fig. 10.19. Scheme of this nuclear transmutation method

The system is summarized in Fig. 10.19. We can obtain the energy and cost balance, which means a cost-free transmutation in an ideal case.

From the macroscopic point of view, this system is a kind of energy converter. We generate γ-rays with high efficiency and finally make hydrogen. During this process, the transmutations for long-lived FPs and TRUs are induced by the γ-rays and neutrons. We can expect to balance the energy. Beside this, there is a possibility that the output of the hydrogen cost can balance the initial cost for installing and cost of operating the system with the partitioning and the separation in this way.

10.5 Conclusions

We have investigated a new method to generate high-brightness γ-rays with high efficiency. This method uses an enhancement of Compton scattering in the photon storage cavity. Applications of the γ-rays for the transmutation is proposed and investigated. Advantages of this method are as follows:

1. energy and cost balance can be achieved;
2. fast transmutation is possible;
3. the system can be shut down immediately and can result in additional safety;
4. the compactness of the electron beam storage ring and CW laser imply low cost of total system; and
5. not far from present-day technology.

We performed laser photon storage and interaction experiments with low-energy electron beams in the cavity. The results corresponded quite well to the predictions for the cavity storage rate and electron beam energy. The preliminary transmutation experiments were performed to determine the reaction rate. Experiments for measuring the neutron energy spectrum and for the enhanced coupling are under way.

Some outstanding issues are as follows:

1. high-power electron beam storage ring,
2. photon storage cavity unit and high-power single mode lasers,
3. electron orbit in the storage ring by multiple Compton scattering,
4. neutron spectrum from the first target to obtain energy balance, and
5. target interaction and reaction rate.

These items are under investigation.

We can generate high-bright γ-rays with high efficiency and use these for transmutation. Energy balance can be expected. Hydrogen production can offset initial and operating costs of the system.

References

1. K. Imasaki: JPN PAT 2528622 (1994), J. Chen, K. Imasaki: Nucl. Instr. Meth. A **341**, 346 (1994)
2. K. Imasaki: The Rev. of Laser Engin. **27**, 14 (1999)
3. K. Imasaki, A. Moon: SPIE **3**(886), 721 (2000)
4. A. Moon and K. Imasaki: J. Jpn. Soc. IR Sci. Tech. **8**, 114 (1998)
5. A. Moon and K. Imasaki: Jpn. J. Appl. Phys. **3**(8), 2794 (1998)
6. M. Nomura et al.: JNC Tech. Rep. JNC TN **9**(410), 2000 (2000)
7. D. Li, K. Imasaki, M. Aoki: J. Nucl. Sci. Tech. **3**(9), 1247 (2002)
8. M. Harakeh and A. von der Woude: *Giant Resonance* (Oxford University Press, Oxford, 2001)
9. A. Moon, K. Imasaki: Rev. Laser Engin. **2**(6), 696 (1998)

10. W.R. Nelson, H. Hirayama, W.O. Roger: *The EGS4 Code System*, SLAC-Report, (1985) p. 265
11. D. Li, K. Imasaki, S. Miyamoto, S. Amano, T. Mochizuki: Rev. Laser Engin. **32**, 211 (2004)
12. D. Li, K. Imasaki, S. Miyamoto, S. Amano, T. Mochizuki: Nucl. Instr. Meth. A **528**, 516 (2004)

Potential Role of Lasers for Sustainable Fission Energy Production and Transmutation of Nuclear Waste

C.D. Bowman[1] and J. Magill[2]

[1] ADNA Corporation, 1045 Los Pueblos, Los Alamos, NM 87544, USA
   Cbowman@cybermesa.com
[2] European Commission, Joint Research Centre, Institute for Transuranium
   Elements, Postfach 2340, 76125 Karlsruhe, Germany
   Joseph.Magill@cec.eu.int

Abstract. While means for transmutation of nuclear waste using fast reactor technology and reprocessing have existed for many years, this technology has not been deployed primarily for economic reasons but also owing to safety and proliferation concerns. Geological storage also remains politically uncertain in some countries as a means for disposal of nuclear waste. We argue here that neutrons supplemental to fission neutrons first from accelerators and later from fusion combined with subcritical systems could displace the need for reprocessing at less cost than reprocessing. Nearly all of the actinide and long-lived fission products from today's reactors could be burned away without reprocessing and the full uranium and thorium resource, which is a greater energy resource than lithium-based d–t fusion, could also be exploited with concurrent burning of the waste. It is shown that a laser–fusion system driving a subcritical fission system and operating at physics breakeven with the recirculation of 10% of the fission electric power would match today's accelerator–spallation technology as a subcritical fission driver and that a fusion system operating at engineering breakeven for driving a subcritical fission system probably exceeds the potential best performance of any known accelerator technology. This chapter advocates an innovative reactor technology beyond those envisaged 50 years ago that still dominate the field. It also calls for a focus of fusion research on fusion neutron production in addition to fusion energy as it shows that fusion-neutron–driven fission should reach technical and economic practicality long before the smaller resource of pure d–t fusion energy becomes practical.

11.1 Introduction

As the world's need for energy grows, the full benefit of nuclear energy still seems to be out of reach. Practical fusion energy has yet to be realized. Existing light water reactors produce a radioactive waste stream that cannot be economically transmuted at present, and geologic storage remains a politically

C.D. Bowman and J. Magill: *Potential Role of Lasers for Sustainable Fission Energy Production and Transmutation of Nuclear Waste*, Lect. Notes Phys. **694**, 169–189 (2006)
www.springerlink.com

charged issue worldwide. Fast reactors could provide access to the full uranium and thorium resources and contribute to waste burning from the LWRs but are more expensive than LWRs. Therefore, although the fast reactor and reprocessing technology has existed for decades in the United States, Europe, Russia, and Japan, the technology has not been deployed primarily owing to cost but also to safety and proliferation issues. The nuclear waste issue remains unsolved.

The purpose of this workshop is to draw attention to laser-driven nuclear reactions that can be induced both by laser fusion class lasers operating at the one megajoule level [1] and by much smaller "tabletop" lasers with much smaller energy in the pulse but much higher power arising from ultrashort pulses. Such small lasers are able to accelerate ions to energies of several hundred MeV in a few millimetres with the obvious capability to induce nuclear reactions other than neutron production. Papers in this workshop describe experiments that have transmuted long-lived unstable nuclei to stable nuclei [2], raising the question of whether this approach might ultimately become a practical means for destruction of the long-lived species in spent reactor fuel [3]. This new prospect must be considered not only in the light of existing reactor technology but also in regard to other technology that might become available. The role of chemical reprocessing and supplemental neutrons to those from fission are central to these questions.

The problem of inadequate fission neutrons and the issues of cost, safety, and proliferation have fostered a complexity for nuclear energy systems that considerably weakens their intrinsic advantage over competitors in terms of fuel cost and energy density. Figure 11.1 shows the simple process steps involved for coal, gas, and wind. Figure 11.2 shows a typical infrastructure for an advanced "double strata" fuel cycle. In this particular fuel cycle there are three reactor technologies, two reprocessing technologies, two fuel fabrication technologies, 21 transport operations in tracing the fuel and waste from beginning

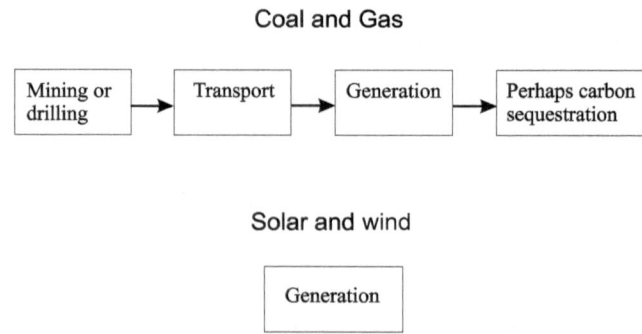

Fig. 11.1. The basic infrastructure elements required for coal, gas, solar, and wind are shown for comparison with the infrastructure for future nuclear power systems shown in Fig. 11.2

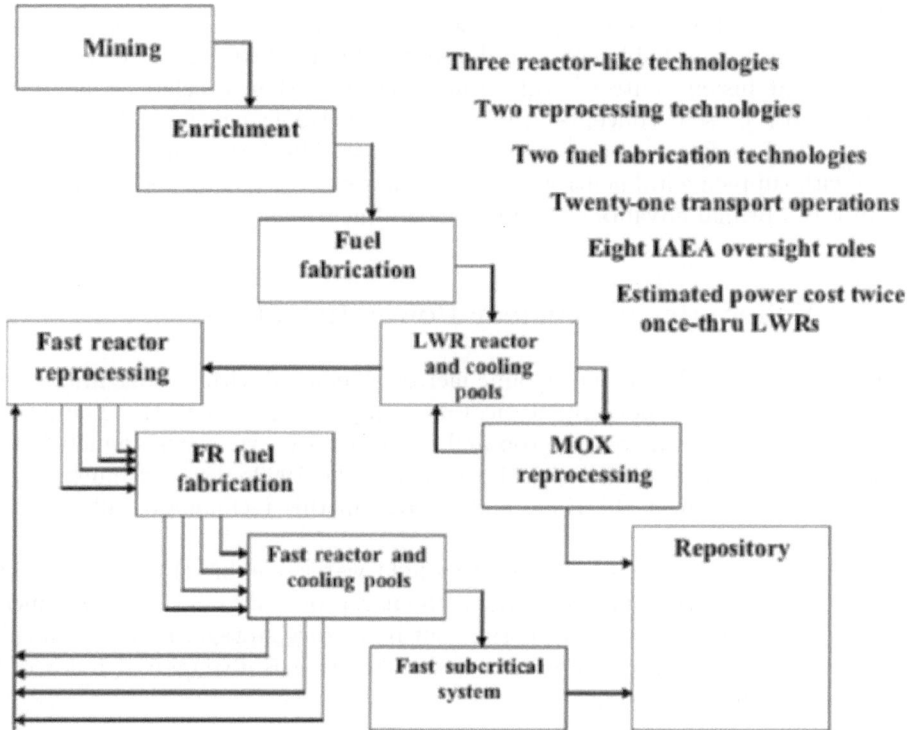

Fig. 11.2. Typical complex advanced "double strata" fuel cycle proposed as a future nuclear power system

to end, and eight IAEA oversight roles. It is not surprising that the cost of this system is greater than the once-through system.

As for other options, most of the reactor developments are based on incremental improvements of concepts proposed 50 years ago. New technology must concurrently address the waste, safety, and proliferation concerns without increases to the cost of nuclear energy.

The problems of nuclear power ultimately have their root in too few neutrons per fission. If instead of 2.5 neutrons per fission there were about 20% more neutrons per fission, breeding would be possible without reprocessing and reactors would have sufficient neutrons to burn most of their own waste. Reprocessing, however, is necessary with current nuclear technology to reduce the waste of neutrons to nonbeneficial neutron loss owing mainly to parasitic neutron capture. However reprocessing is a mature, albeit expensive technology. In the early days of nuclear power, the possibility of supplementing the fission neutrons by accelerator sources was far beyond the accelerator technology, but 50 years of accelerator progress have reduced the cost of accelerator-produced neutrons and improved reliability. The choice of fission technology based on reprocessing instead of using supplemental accelerator produced neutrons is no longer so clear and today's cost of accelerator produced neutrons probably can be reduced by a further factor of 2. We will show below that fusion neutrons will be much cheaper than accelerator neutrons long before present concepts of fusion power becomes economically competitive if in fact fusion energy is a realistic goal. The future of nuclear power lies with supplemental neutrons and with new nuclear reactor technology that can take optimal advantage of supplemental neutrons.

11.2 Economics of Nuclear Power Initiatives

An indication of the main economic factors associated with various initiatives (based on systems and technologies) are compared in Table 11.1. The figure shows technologies across the top and a list of initiatives down the left-hand side. The revisiting of subcritical systems in the 1990s received much international attention at the time, and interest in this technology still remains high.

Starting at the top we have existing light water reactors (LWRs) based on solid fuel and a thermal spectrum without reprocessing and with the once-through fuel going directly to permanent repository storage. Power from this system is economically competitive and therefore it is given a cost index of 1 in the right-hand column.

Table 11.1. Breakdown of the economic factors associated with various nuclear power initiatives based on different technologies. In order to be economically competitive, initiatives using supplemental neutron technology and reprocessing may be mutually exclusive

System	Liquid Fuel Thermal Spectrum	Solid Fuel Thermal Spectrum	Fast Spectrum	Accelerator of Fusion Driven	Reprocessing	Power Cost Index
Existing LWRs		X				1
Internal mainstream technology		X	X		X	2
Rubbia'96			X	X	X	3
Europe'04			X	X	X	3
Japan'04			X	X	X	3
Los Almos'94	X			X	X	2.5

The international mainstream of reactor technology development is to include not only the LWRs but also fast reactors for breeding fuel for the LWRs and reprocessing to prepare the fuel for both the LWRs and the fast reactors. Although this approach is based on fully developed technology, it has not yet been deployed. With the introduction of fast reactors and reprocessing, it is given a cost index of 2 in the present comparison.

Rubbia led an effort promoting the introduction of the accelerator to fast spectrum systems that also required reprocessing and this system is still under consideration in laboratories in Europe and Japan. The introduction of an accelerator goes significantly beyond the international mainstream reactor initiative based on fast reactors and reprocessing and, for this reason, is given a cost index of 3.

The Los Alamos accelerator-driven thermal spectrum concept introduced earlier not only had the economic advantage of using a cheaper graphite thermal spectrum technology, but also required supplemental neutrons from an accelerator. Since this system is based on thermal reactor technology as well as on accelerator and reprocessing technologies, the cost index is expected to be between 2 and 3 as shown in Table 11.1.

Following the breakdown given in Table 11.1, initiatives based on the use of subcritical systems requiring supplemental neutrons have an associated economic penalty as do systems involving reprocessing. The introduction of *both* supplemental neutrons and reprocessing increases the economic burden further. It seems clear that economic practicality eliminates the inclusion of both reprocessing and supplemental neutrons for economically viable nuclear technology. The initiative advanced here is to include supplemental neutrons while eliminating reprocessing.

11.3 Technology Features for New Initiatives

From the investigations of transmutation technologies over the past 15 years, the following observations can be made.

First, the often-stated advantages of the fast spectrum over the thermal spectrum are questionable. The neutron economic advantage arising from the fast spectrum capability to fission both even and odd mass nuclei is real but overstated because the thermal spectrum wastes fewer neutrons on fission products and operates with less leakage. The thermal spectrum also has significant advantages in lower cost components and higher transmutation performance [4] with less proliferation concerns associated with its much smaller inventory. Another often-quoted advantage of the fast spectrum is less build-up of higher actinides of americium and curium that greatly complicate reprocessing. Obviously if the technology minimizes or even avoids reprocessing, as will be shown below, this point also is irrelevant.

Second, the complications of solid fuels – reactivity swings in going from fresh to spent fuel, limits to fuel lifetime, downtime for fuel changes, and the expense of fuel fabrication – can be avoided with liquid fuels (molten salt reactors are one of the reactor systems considered within Generation IV). The choice of molten fluoride salt as the fuel medium enables operation in the 750°C temperature range without the expense and safety concerns related to a high-pressure containment vessel, and it enables a high thermal-to-electric conversion efficiency. By designing for low power density and flow of the fuel against the containment vessel wall, it is practical for the vessel to act as the heat exchanger. Liquid fuel never leaves the vessel until the end of life of the reactor. The flow of liquid fuel through the system avoids the problem of uneven burnup of fuel in solid fuel systems. In the case of loss of coolant or loss of powered fuel flow, the natural convection carries the fuel by the containment wall allowing heat removal by natural means while the negative temperature coefficient of the liquid fuel automatically controls the chain reaction. The transfer of fuel at end of reactor life is readily accomplished by means of helium-pressurized piping to storage canisters without the expensive mechanical fuel removal and transfer systems required by solid fuel systems.

Third, graphite as a moderator has been well established by the Molten Salt Reactor Experiment (MSRE) at ORNL in the 1960s as a stable material that is fully compatible with molten salt. Graphite is much cheaper than the steels used in fast reactors and LWRs, and the system can also include substantial granular graphite. The cost of granular graphite is less than 10% of solid graphite. Moreover, the ADNA collaboration has discovered that modern graphite can be cheaply produced with energy stored in the graphite lattice and that this energy can be transferred to the neutrons [5]. The important consequence of stored energy and energy transfer to the neutrons is that room temperature graphite establishes an average temperature of the neutrons of about 2000 K. At this neutron temperature the neutron spectrum overlaps favorably with resonances near 0.3 eV in ^{239}Pu, ^{241}Pu, ^{237}Np, ^{241}Am, and ^{242m}Am. The reaction rates, with higher actinide fuel, of neutrons in "hot" graphite systems are therefore much higher than those in classical graphite systems. In addition, parasitic losses of neutrons are a factor of 3 lower than at room temperature and "hot" graphite reflector performance is a factor of 2 better. Therefore, the new "hot" graphite offers major favorable consequences for system performance with associated capital and operations cost reductions that may be added to those reductions arising from the use of liquid fuel. It seems likely that a thermal-spectrum molten-salt "hot"-graphite reactor will produce power at well below the total cost of today's once-through LWR.

11.4 The Sealed Continuous Flow Reactor

The sealed continuous flow reactor (SCFR) with recycle, shown in Fig. 11.3, is one concept that avoids the fast neutron spectrum and the use of solid fuels and introduces that advantages of graphite described above. It is constructed

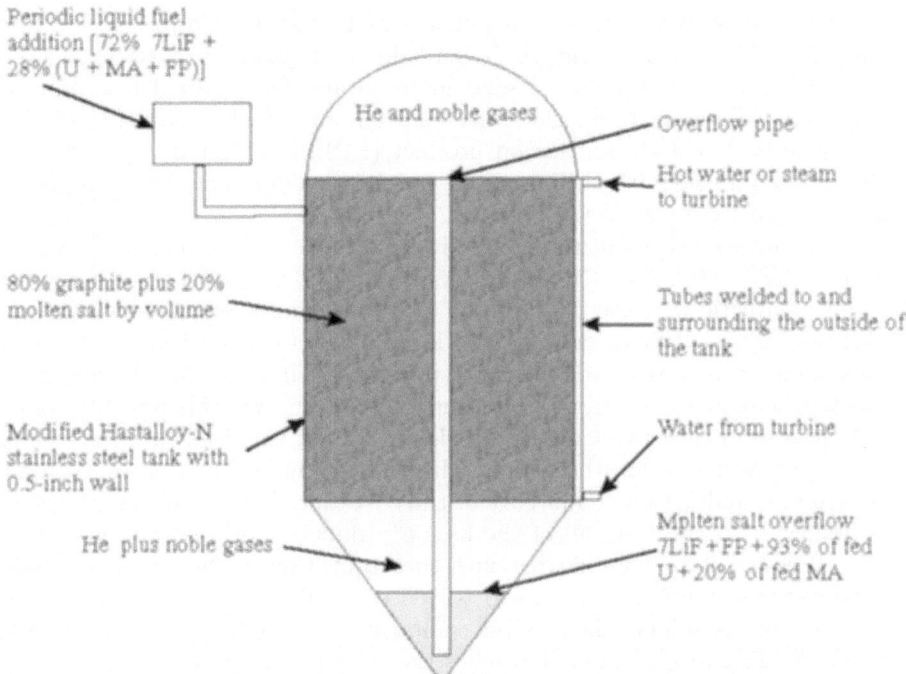

Periodic liquid fuel
addition [72% 7LiF +
28% (U + MA + FP)]

He and noble gases

Overflow pipe

Hot water or steam
to turbine

80% graphite plus 20%
molten salt by volume

Tubes welded to and
surrounding the outside of
the tank

Modified Hastalloy-N
stainless steel tank with
0.5-inch wall

Water from turbine

He plus noble gases

Mplten salt overflow
7LiF + FP + 93% of fed
U + 20% of fed MA

Fig. 11.3. This schematic drawing of a thermal spectrum molten salt reactor is shown with a central critical volume of 80% granular graphite and 20% molten salt fuel. The reactor in the presence of a large amount of ^{238}U has strong negative feedback and need not have control rods but only scram rods. An overflow pipe on the central axis keeps the fuel level the same as fuel is added as shown from the *upper left-hand* side at the rate of perhaps 2 litres per day. Heat is removed by forced flow of fluid up through the center and down beside the outer wall of the tank that has 2-cm-diameter pressure tubes with water entering at the bottom for driving a steam turbine for electricity production. Volumes *top* and *bottom* are provided for collection and storage of noble gases, and they may have storage capacity sufficient for the life of the reactor. The system is a sealed unit never opened even during filling as filling may be accomplished with a two-stage process or even three stages if necessary. All materials in this system were shown by the MSRE program at ORNL in the 1960s to be fully compatible with molten salt, so the materials development for this reactor was completed years ago

as a tank filled with granular graphite and with a molten fluoride salt consisting of a ^{7}LiF carrier mixed with the fluoridized constituents of commercial reactor spent fuel. Fission energy is generated in the salt, which undergoes pumped circulation upward in the center and downward by the outside wall. Heat is transferred through the steel wall to tubes welded to the outside of the steel tank where water is heated to high temperature for driving a steam turbine for high-efficiency electric power generation. To produce the fluoridized spent fuel, commercial LWR spent fuel is first exposed in a plasma torch facility to chlorine gas to convert the zircalloy fuel cladding to $ZrCl_4$ gas. The remnant oxide fuel remains as solid material and is converted in a fluidized bed chemical reactor to fluoride salt. Thus all of the fuel including the U, Pu, minor actinide (MA), and fission product (FP) is converted to fluoride salt that is then mixed with ^7LiF, melted, and fed as a granular solid fluoride salt or as a eutectic liquid to the reactor.

The fluoride salt mixture from LWR spent fuel contains in mole percentage about 92.5% ^{238}U, 2% ^{239}Pu and MA, 1.5% ^{235}U, and 4% FP. This mixture forms a eutectic with melting point 550°C when mixed in 3:1 mole ratio with ^7LiF and about 2 litres or about 7.5 kg of spent fuel is fed into the system once a day and mixes with the other molten salt in the tank. As the fuel in the tank begins to rise from adding new fuel, overflow spills into the conical holding tank below through the overflow tube. At this feed rate the average atom spends 4 years in the neutron flux, 7% of the fed ^{238}U is converted to plutonium, and fissioned and 80% of the ^{235}U and ^{239}Pu and minor actinide is fissioned. In addition, all of the fission products, both fed and generated, absorb neutrons to some degree with the absorption in the long-lived fission products transmuting them to short-lived or stable nuclides.

A comparison [6] of the isotopic composition of the Pu and minor actinide before feeding to that after it reaches the storage tank is shown in Fig. 11.4. The isotopic feed distribution is shown as the back column with the fraction of all isotopes adding to 1.0 and with $^{239,241}Pu$ being the major constituent at about 58%. The isotopic composition in the storage tank after burning is shown at the front and the total adds to 0.212 indicating about 80% of the Pu and minor actinide was burned. The $^{239,241}Pu$ content is reduced by more than a factor of 10 in a single pass.

The fraction of burnup of the Pu and minor actinide in a fast spectrum in a single pass with fluence limited by the reduction in reactivity of the solid fuel as the plutonium burns away is shown in the middle for comparison. It is seen that the total actinide burnup is $1.0/0.67 = 1.5$ instead of 5 for the thermal spectrum and that the fraction of $^{239,241}Pu$ content is reduced only by a factor of 2 instead of 10 for once through the thermal spectrum. It should be noted that the performance of this graphite reactor will be significantly further enhanced with the use of the "hot" graphite referred to earlier.

This molten salt reactor must be brought to isotopic equilibrium after start-up with a molten salt mixture with the same reactivity as the overflow salt. Full equilibrium is then established in a couple of years. In equilibrium, it is to be noted that the salt in the overflow tank is exactly the same chemical and isotopic composition as the salt in the neutron flux above. Thus when sufficient salt has been accumulated over about 4 years in the overflow tank to equal the volume in the neutron flux above, the accumulated overflow salt may be used to start an identical reactor fed with the same input as the first, except that the second reactor is in equilibrium from the beginning. An

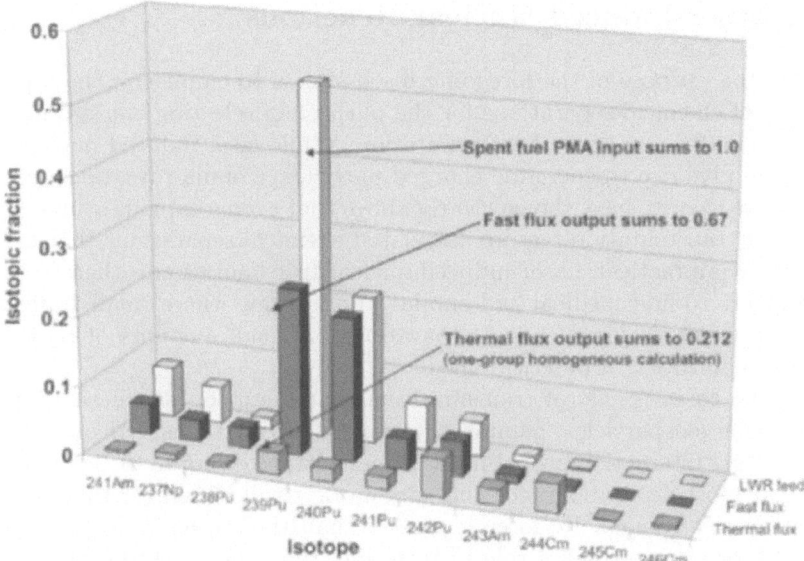

Fig. 11.4. The isotopic distribution of the plutonium and minor actinide (PMA) in the spent fuel from an LWR at 30,000 megawatt-days per ton burnup is shown at the back of the figure and the columns sum to 1.0. The isotopic distribution at the front is that of the overflow material in the reactor of Fig. 11.3 and the columns sum to 0.212 indicating that the total fed material has been reduced almost by a factor of 5. The distribution in the middle shows the burnup achieved in one pass through a fast spectrum reactor for comparison. The total reduction is only by a factor of about 1.5 with less effective burnup of the fissile isotopes ^{239,241}Pu. The products from the burnup of any uranium present are not included in this figure but the distribution of the nonuranium isotopes is little different from that shown at the front of the figure

original reactor therefore may be the mother of about ten other reactors over a lifetime of about 40 years.

It is further worth noting that the salt fed into the tank at the top is converted immediately after mixing to the same composition as the salt in the neutron flux (and the salt in the overflow tank). This is a significant nonproliferation advantage offered by this type of liquid fuel system because in a solid fuel reactor, the fuel added in the presence of inspectors could be removed later with little or partial burnup and potentially diverted for weapons use. Once the fuel has been added to this reactor, it is immediately converted to its final isotopic distribution.

11.5 Laser-Induced Nuclear Reactions

Part of the purpose of the foregoing discussion is to emphasize the economic penalty of chemical separations for the purpose of selecting out specific long-lived nuclei for any type of transmutation. While laser-induced nuclear reactions can give rise to energetic charged particles, gamma rays, and neutrons, we believe that we have shown clearly above that even a copious source of such agents of transmutation require associated chemical separations that are economically impractical. Laser-induced nuclear transmutation is therefore much more likely to find medical and similar applications where small transmuted quantities and simple chemical separations on small amounts of radioactive isotopes can satisfy a practical need.

Of the three agents of transmutation that might be produced by pulsed lasers (charged particles, gamma rays, and neutrons), charged particles will have weaker interaction probabilities than neutrons owing to the coulomb barrier that must be surmounted, and gamma rays will be weaker than neutrons since the electromagnetic force is weaker than the strong force. Therefore if pulsed lasers are to play a role in transmutation, they must be designed for copious neutron production.

With regard to laser-produced neutrons, the options can be divided between endoergic reactions and exoergic reactions. Exoergic reactions offer the advantage that the neutrons may have a sufficient energy to multiply themselves by subsequent (n,2n) reactions and the charged particle accompanying the neutron production might deposit its energy so as to promote further neutron production in a target raised to high temperature by the laser. Endoergic reactions such as ^{7}Li(p,n) reactions offer neither high-energy neutrons nor significant energy associated with the charged particle. If the choice then must be for exoergic reactions that produce neutrons with as associated charged particle to further heat the neutron production medium, the (d,t) reaction is hard to beat. The route to transmutation with high-power lasers must therefore be through neutrons produced via the (d,t) reaction operating on waste from today's power reactors without prior chemical separations.

11.6 Introducing Fusion Neutrons into Waste Transmutation

The reactor shown in Fig. 11.3 is an effective transmuter of plutonium and minor actinides and effective also in burning 7% of the fed ^{238}U with neutrons produced only by fission. To burn the exit stream further will require external neutrons to supplement the fission neutrons and today these can be most effectively produced by an accelerator. This reactor design can accommodate a source of neutrons by placing the neutron production target near the center. This might be done by transporting the accelerator beam or beams into the center through one or more tubes, or the reactor might be split in two to allow a 30- to 100-cm wide space for a target to be placed between the two halves. With either adaptation for the target for neutron production, the exit stream from the SCFR could be fed into a reactor of the same SCFR design but with a (d,t) source and an effective multiplication of $k_{eff} = 0.96$ instead of the value of $k_{eff} = 1.0$ for a reactor. In that case the reactor exit stream could be burned further with about 7% more of the ^{238}U being fissioned so that the new exit stream would contain about 86% of the original ^{238}U. In addition, the plutonium, minor actinide, and fission product from the LWR and that produced in the first SCFR burn stage would be further burned away.

Accelerator technology is fairly mature and it seems unlikely that neutron production costs can be reduced further by more than a factor of 2. However, if the cost of accelerator produced neutrons can be reduced by a factor of 2 beyond today's technology, the second exit stream could be recycled again with $k_{eff} = 0.92$ and the ^{238}U burned down to about 80% with the other transmutation benefits to plutonium, minor actinide, and fission products. If fusion neutrons in the future are produced even more cheaply than from the most mature accelerator technology, then recycle without reprocessing can continue further.

Figure 11.5 shows that d–t fusion neutrons must be much cheaper than accelerator neutrons if an economically practical fusion power plant is a realistic objective. It is based on the practical assumption that the capital cost per electric kilowatt of the fusion power plant can be no more than that of fission reactors achieving a 45% thermal-to-electric conversion efficiency, that the fusion reactors will achieve the same thermal-to-electric conversion efficiency, and that the operations cost will also be equivalent. With these assumptions the cost of fusion electric power will be about the same level as today's high thermal-to-electric conversion efficiency fission electric power plants.

In order to place the fusion-neutron–driven fission reactors on the same footing as the accelerator-driven fission system, we first find the percentage of the electric power generated in a fission system that must be used to drive the accelerator (see the Appendix). The reactor of Fig. 11.3 would generate 100 MWe power operating at a fission power of 222 MWt corresponding to a thermal-to-electric conversion efficiency of 45%, which seems practical at its 750°C operating temperature. At 200 MeV per fission this corresponds to 7×10^{18} fissions per second.

For a subcritical system, power is generated by many fission chains of finite length rather than in a continuous chain. The length or the average number of fissions events per fission chain is $1/(1 - k_{eff})$, where k_{eff} is the effective multiplication constant; k_{eff} is 1.0 for a critical reactor and less than 1.0 for subcritical reactors. The finite fission chains obviously must be started by a neutron, but not all neutrons injected into a subcritical system will start a fission chain. If injected neutrons are absorbed like fission neutrons and on average neutrons are produced in each fission, then only the fraction k_{eff}/ν of the neutrons injected start fission chains. The fraction might be larger than this if the injected neutrons can be made to be more effective in starting fission

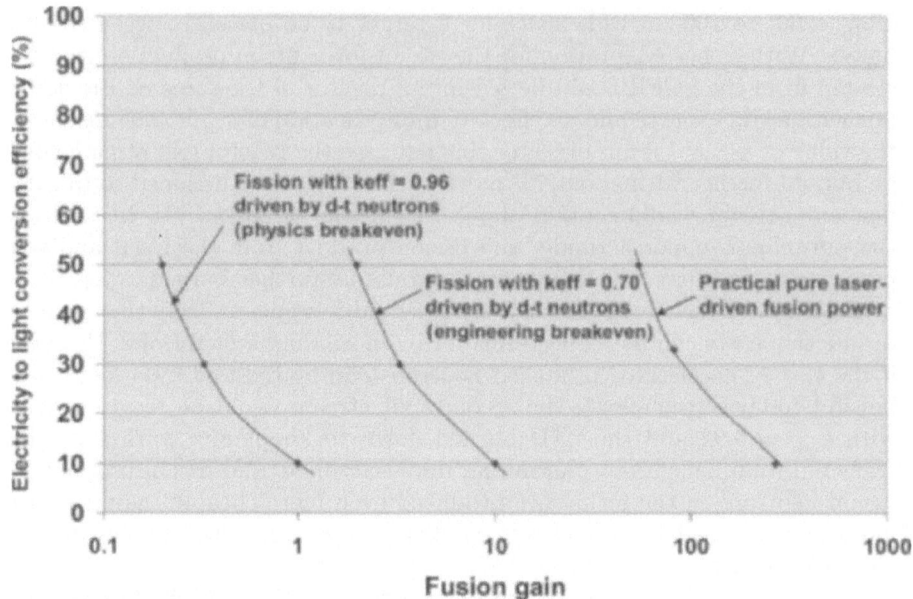

Fig. 11.5. Fusion research objectives and related neutron production significance; electricity-to-light conversion efficiency vs. gain. When the fusion gain is 1.0 (physics breakeven) and the conversion efficiency of electricity to light is 10% and about 8.5% of the 100 MWe fission electric energy is needed to drive the laser, inertial d–t fusion as a neutron source matches the performance of today's practical spallation source (see the Appendix). This assumes a 1 GeV accelerator using 8.9 MWe to produce 4 mA of proton beam. The neutron intensity from the accelerator or the laser is sufficient to drive a fission system at a power of 222 MWt at $k_{eff} = 0.96$. The curve on the left also shows that equivalency with the accelerator may also be reached with a 50% conversion efficiency and a gain of 0.20. The middle curve shows that the system may be driven at the same fission power with $k_{eff} = 0.70$ for engineering breakeven performance of laser fusion. The curve on the right shows the efficiency vs. gain for a practical pure laser fusion device with 13.8% of the fusion electric power used to drive the laser. Clearly laser fusion as a neutron source is highly effective as a fission subcritical driver allowing access to effectively an "infinite" fission energy source long before the pure laser fusion power goal is reached

chains than the fission neutrons, but in general this significantly complicates the geometry of the subcritical reactor and we assume here that fission and injected neutrons are equivalent. The average number of fissions started by a single injected neutron therefore is $k_{eff}/[(1-k_{eff})]$. For $k_{eff} = 0.96$ for a typical spallation-driven system and an average number of neutrons per fission of $\nu = 2.5$, the number of fissions per injected neutron is found to be 9.6. Therefore the neutron injection rate for a 100 MWe system is $(7 \times 10^{18})/9.6 = 7.3 \times 10^{17}$ per second. A 1 GeV proton on a lead target produces by spallation about 30 neutrons per proton so the number of protons required is 2.4×10^{16} per second.

This corresponds to 4.0 mA or 4 MW of proton beam power requiring 8.9 MW of electric power for a 45% efficient accelerator. Such an accelerator seems well within today's linac technology as the 100 MW beam power accelerator proposed for the Los Alamos Accelerator Production of Tritium (APT) project was found to be practical from both technical and a targeting perspectives. The cost of a 4 MW accelerator with production line manufacturing might be about $125 million and probably could be economically practical selling power at the competitive cost of $0.04/kilowatt hour. The primary point of this chapter is to determine the performance characteristics of a d–t fusion source that would perform the equivalent function of an accelerator spallation neutron source (see the Appendix).

A subcritical reactor of 100 MWe capacity driven by a d–t fusion source with $k_{eff} = 0.96$ would require the same number of neutrons as the accelerator-driven system except that a 14.1 MeV fusion neutron can be multiplied by a factor of about 2.5 by a surrounding blanket of beryllium or lead. The number of fusion neutrons required is therefore $7.3 \times 10^{17}/2.5 = 2.9 \times 10^{17} n/s$. With each fusion event producing 17.6 Mev, the fusion thermal power generated is 0.82 MW. If the fusion gain is 1.0(physics breakeven), and the laser coverts electric power to laser power with an efficiency of 10%, the electric power required by the laser is 8.2 MW or 8.2% of the 100 MWe generating power capacity. The 100 MWe could be obtained from a laser operating with a gain of 0.3 and an efficiency of 33% or a gain of 0.2 and an efficiency of 50%. This is shown by the curved line on the left of Fig. 11.5. Performance at least reaching physics breakeven (although at a very low pulse rate) is expected from the National Ignition Facility (NIF) soon to come on line at the Lawrence Livermore National Laboratory. So a fusion neutron source operating at physics breakeven (a fusion gain of 1.0), at an electricity-to-light conversion efficiency of 0.1, and at an adequate pulse rate would match the neutron generation performance of the best of today's accelerator technology.

The next goal for fusion on the road to practical fusion power is engineering breakeven when the fusion power equals the electric power needed to drive the laser. For an electricity-to-light conversion efficiency of 10%, this is achieved when the gain reaches ten (or 3.33 for 30% efficiency or 2 for a 50% conversion efficiency, or 100 for a 1% efficiency). One may follow the same process for calculating the performance of laser-driven fission power and find that a subcritical liquid fuel fission power system operating at $k_{eff} = 0.70$ then becomes economically practical as shown in the middle of Fig. 11.5. It seems unlikely that relatively mature accelerator technology will provide neutrons at a cost lower by a factor of 2 than present systems and so the limit for accelerator-driven systems is probably k_{eff} no lower than 0.92. Thus, long before one reaches practical pure fusion power, fusion engineering breakeven would enable practical subcritical fission systems operating at $k_{eff} = 0.70$. Further improvements in gain by a factor of 2.5 would enable subcritical fission systems with $k_{eff} = 0.50$. Such a system as shown in Fig. 11.3 would enable the burnup of half of the world's uranium and thorium with concurrent burnup of nearly all of the plutonium, higher actinide, and long-lived fission product without reprocessing.

Using the same line of reasoning, one finds that a practical pure fusion system operating at the same electric power output as a fission system, with the same capital and operating cost as the fission system, achieving a 45% thermal-to-electric conversion efficiency, and with 8.2% of the generated electric power being fed to the laser, requires a gain of about 270 for an electricity-to-light conversion efficiency of 10%. This is another factor of 27 performance improvement beyond engineering breakeven as shown on the right-hand side of Fig. 11.5.

One sees therefore that with an electricity-to-laser light conversion efficiency of 10%, fusion neutrons are economically competitive with accelerator neutrons at physics breakeven for a gain of 1. This gain is a factor of 270 lower than required for pure fusion power. At engineering breakeven, equivalent to a gain of 10, fusion neutron sources become economically practical drivers for subcritical systems operating at $k_{\text{eff}} = 0.70$. Under this condition, about half of the energy resource from Th and U can be recovered in liquid fueled systems without reprocessing and while consuming nearly all of the plutonium, minor actinide, and long-lived fission product generated.

11.7 Comparison of the Fission and d–t Fusion Energy Resources

Half of the U and Th fission energy resource is an enormous amount of energy even exceeding the energy resource from d–t fusion. ^{6}Li is the source of tritium for d–t fusion and its isotopic abundance [7] in mg/kg is about 52nd among the isotopes compared with 47th for ^{238}U and 38th for ^{232}Th. Converting to moles instead of milligrams and taking 17.6 MeV for fusion energy reaction and 200 MeV per fission, we find that the average energy density in the earth's crust is about 400 MJ/kg from d–t fusion, 200 MJ/kg from ^{238}U, and 800 MJ/kg from ^{232}Th. Even with burning only half of the Th and U, the fission energy resource still exceeds that from fusion [8].

As for accessing the U, Th, or ^{6}Li, Cohen [9] gives the cost for extracting uranium from the ocean as \$250/lb, which is about ten times the cost of mined uranium. However, present commercial power plants extract only about 1% of the available energy from mined uranium although the system described here with accelerator- and fusion-driven recycle would extract 50 times as much energy. Cohen also estimates that the amount of uranium in seawater would supply the world's current electricity usage for 7 million years and the thorium would add 28 million more years. He also points out that seawater uranium levels are being replenished by rivers that carry uranium out of dissolved rock at a rate sufficient to provide 20 times the world's current total electricity usage and that this process could continue for a billion years.

11.8 Implications for Fusion Energy Research

First, the fission energy resource exceeds that from d–t fusion energy and the fission resource far exceeds the world's energy needs on the timescale of many millions of years. From the energy resource perspective, fusion energy is unnecessary if fission energy can be accessed as described here, but fusion neutrons might be necessary for driving subcritical fission systems to enable access to the full fission resource with concurrent burnup of most of the undesirable fission by-product species.

Second, fusion research becomes economically deployable as a neutron source for driving fission systems when it reaches scientific breakeven at an adequate pulse rate with an electricity-to-light conversion, efficiency of about 10% and when about 8.5% of the fission electric power is consumed in driving the laser. Reaching this gain of 1 with an electricity-to-laser light conversion, efficiency of 10% would match the performance of the best of known accelerator technology for spallation neutron production. The NIF facility [10] might reach scientific breakeven in a few years. With a next objective of reaching scientific breakeven while consuming only about 10% of the fission power produced by the fusion neutrons, laser fusion would be competitive with the best of today's mature accelerator technology as a subcritical fission reactor driver.

Third, pushing fusion neutron source development another factor of ten to engineering breakeven would enable the burning of about 40% of the ^{238}U in commercial reactor waste while burning without reprocessing nearly all of the plutonium, minor actinide, and long-lived fission product. More important, it would enable the exploitation of both the thorium and the uranium resource recovered from seawater at fuel cost rates far lower than the present costs of mined uranium. The energy resource accessed is far more than humans can contemplate.

Fourth, fusion technology would have to be pushed another factor of 27 beyond engineering breakeven before it became economically competitive with fusion-neutron–driven fission energy at which point it would only add to an already "infinite" amount of accessible fission energy. The only justification for pure fusion energy after such fission systems are operating would be to claim it as cheaper or cleaner or safer than fission energy. All three justifications seem weak in view of the advantages of the system of Fig. 11.3 probably achieving lower cost than present power from any source, much reduced long-term radioactivity, and much safer operation owing to subcriticality, the absence of reprocessing, and the virtual elimination of fuel transport in the fuel cycle. It appears that effort to proceed beyond engineering breakeven to competitive pure laser–fusion power by improvement by a factor of 27 is unlikely to be rewarding from an energy resource perspective.

Fifth, the above arguments apply to both inertial fusion and magnetic fusion approaches, but inertial fusion allows a bulky laser light source to be placed far from and outside of the fission region. Presently the dominant magnetic fusion approaches require that the fission reactor and fusion neutron production system occupy the same volume creating perhaps unsolvable engineering problems. Magnetic fusion energy therefore seems useful only if and when it becomes an economically competitive pure fusion energy source. However, inertial fusion has a major application as a neutron source before traveling far down the long road to commercial fusion power.

Sixth, Fig. 11.6 shows the greatly simplified infrastructure of recycling liquid fuel in a thermal-spectrum without reprocessing but with supplemental neutrons first from an accelerator and later from inertial fusion. It is almost as simple as Fig. 11.1 and provides access to the full nuclear energy resource and means for destruction of the waste from today's reactors with 3 transport steps instead of the 21 of Fig. 11.2. In addition, this technology can be implemented without actinide enrichment, or the production of weapons-useful material, or reprocessing, so it has strong nonproliferation advantages. Although the development of the supplemental neutron sources is not trivial, we believe that this approach may provide an alternative to reprocessing, and be economically competitive with any nonnuclear energy technology except hydropower.

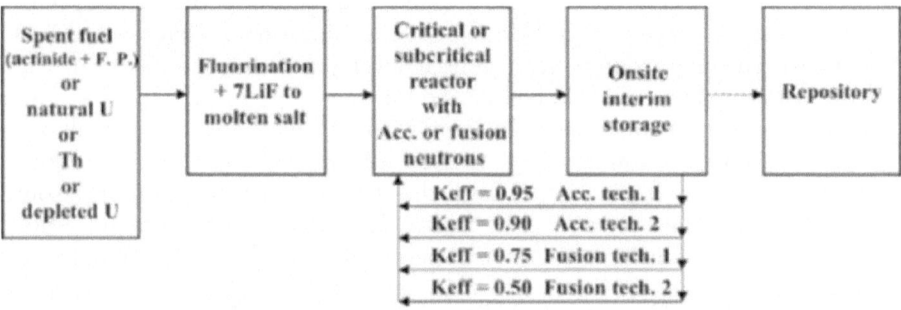

Fig. 11.6. Infrastructure for graphite-moderated thermal-spectrum liquid-fuel critical and subcritical technology. The molten-fluoride salt can accommodate any feed material so spent fuel can be fed including its ^{238}U, all other actinides, and fission products. In addition, the system can accept natural uranium, or thorium, or depleted uranium or uranium and thorium together. The feed material is converted to fluoride salt and fed continuously first to a critical reactor and removed continuously from the critical reactor and stored on site until the material is recycled into an accelerator-driven system. It is further recycled in accelerator or fusion-driven systems until $k_{\mathrm{eff}} = 0.50$ at which point almost half of the fission energy has been extracted from the U or Th and nearly all of the plutonium and minor actinides and long-lived fission products have been eliminated without reprocessing

11.9 Summary and Conclusions – Implications for Nuclear Power R&D

The key elements of this proposed technology are the liquid fuel which allows indefinite recycling without reprocessing, a thermal spectrum that permits small inventories and high single pass burnups, and low-cost neutrons from accelerators and later from (d,t) fusion. Here then are the elements needed with some comments.

- *Thermal-spectrum molten salt reactor revival.* The Molten Salt Reactor Experiment (MSRE) at Oak Ridge successfully demonstrated in the 1960s the molten salt materials technology. It failed owing to the requirement that the fuel be passed through external heat exchangers with the vulnerability that a pipe break from an earthquake might dump large amounts of fuel onto the floor, the requirement for an online reprocessing system, and the fact that it was less effective at breeding plutonium than a fast-spectrum reactor. A new low-power molten salt reactor is needed to demonstrate the recovery of this materials technology.
- *LWR spent fuel fluorination.* Since no chemical separations are required, the removal of the fuel cladding and the conversion of oxide to fluoride is a much simpler process than reprocessing. Logistically, it would be a great advantage if this conversion could be done in a small facility at the reactor site. In any case production-scale fluorination must be demonstrated. Of course if the starting fuel is natural U or Th, no development is necessary.
- *^7Li isotopic separation.* Although enough ^{7}LiF is on hand now for demonstrations, substantial quantities of ^{7}LiF are required for the SCFR technology. Fortunately, isotopic separations for light isotopes is simple and inexpensive compared to that for uranium, but production-scale operations will be required.
- *(d,t) fusion neutron sources.* The sooner practical fusion neutron sources come online the better since fusion offers far more potential "stretch" than accelerator technology. The maximum benefits of the approach outlined in this chapter of full use of the uranium and thorium with transmutation of the waste stream to a high degree comes with the fusion neutron source.
- *Conventional accelerator technology adaptation.* While existing accelerator technology is quite mature for both linacs and cyclotrons, a highly reliable practical design for inexpensive accelerator mass production must be demonstrated. There is time for this and for beam targeting since the first generation SCFRs would be critical systems.
- *Beam targeting and reactor integration.* While spallation is the usually discussed means for accelerator production of neutrons, a light element target producing fewer but higher energy neutrons surrounded by a lead multiplier appears to work as well as spallation. Choosing the optimal target and integrating it into the SCFR requires both conceptual design and demonstration.

- *Advanced accelerator technology development.* While accelerator technology is rather mature, it would be short-sighted to expect that further development could not reduce the cost of accelerator neutrons by a factor of 2 from today's accelerators. Such an accelerator would enable recycle of the fuel from a previous cycle of accelerator-driven operation with very significant contributions to the world's energy supply and would enable additional time, if necessary, for implementation of (d,t) fusion neutron sources.

- *Ultimate disposition of the residual waste.* The liquid fuel output from the SCFR can be stored on the power production site for long periods while waiting for recycle with inventories much smaller than from the LWRs since the SCFR generates so much more energy from a given amount of mined actinide. This recycling process with successively better supplemental neutron sources might extend over hundreds of years so that permanent waste storage would be far into the future but ultimately there will be waste to be stored. Present geologic storage development and other permanent storage options provide a valuable base for assuring the public that means can be found for dealing with a waste stream nearly devoid of plutonium and higher actinides and with a greatly reduced portion of long-lived fission products.

It should be emphasized that accomplishing this list of R&D efforts is not necessary for deployment of this technology, but only for seeing its full potential. We believe that the SCFR can be built with today's technology without supplemental neutrons from either accelerators or fusion and can exhibit highly effective performance as a reactor for transmuting LWR waste as well as first-generation burning of substantial amounts of ^{232}Th and ^{238}U. The addition of supplemental neutrons from external sources enables a high degree of "stretch" for the molten salt thermal spectrum technology that is unlikely to be matched by solid fuel technology.

References

1. D. Besnard: *The Megajoule Laser: A High Energy Density Physics Facility*, this conference, Lasers & Nuclei, ed by H. Schwoerer, J. Magill, B. Beleites (Springer Verlag, Heidelberg, 2005)
2. J. Magill, J. Galy, T. Zagar: Laser transmutation of nuclear materials. In: *Int. Workshop on Lasers and Nuclei, Application of Ultra-High Intensity Lasers in Nuclear Science*, Karlsruhe, Germany, September 13–15, 2004
3. IAEA: Implications of Partitioning and Transmutation in Radioactive Waste Management, Technical Reports Series No.435, 2005. See also J. Magill et al.: Nucl. Energy **42**, 263–277 (2003)
4. V. Berthou, C. Degueldre, J. Magill: Transmutation characteristics in thermal and fast neutron spectra: Application to americium. J. Nucl. Mater. **320**, 156-162 (2003).

5. C.D. Bowman: Thermal spectrum for nuclear waste burning and energy production. In: *Proc. Int. Conf. Nucl. Data Sci. Techno.* Santa Fe, NM (2004)
6. C.D. Bowman: "Once-through Thermal-Spectrum Accelerator-Driven Waste Destruction Without Reprocessing. Nucl. Technol. **132**, 66–93 (2000)
7. CRC Handbook of Chemistry and Physics, ed by David R. Lide (1992).
8. It is to be noted that the energy liberated in the production of tritium from neutron absorption on ^{6}Li is not included in the power calculations since it might not be practical to convert that energy to electric power depending on the system design.
9. B.L. Cohen: Letter in Physics Today, p. 16 (November 2004) and B.L. Cohen, Am. J. Phys., **51**, 75 (1983)
10. The National Ignition Facility (NIF) nearing completion of construction at the Lawrence Livermore National Laboratory at Livermore, CA, is described elsewhere in the proceedings of this workshop (see [1])

11.10 Appendix

11.10.1 Laser Fusion Power Required to Drive a Subcritical Fission Reactor

1. It is assumed that the fission reactor runs at a thermal power level of $P_{\text{fission,th}} = 222$ MWth. With a thermal-to-electric conversion efficiency of 45%, the electrical power generated is $P_{\text{fission,el}} = 100$ MWe. The fission rate required to sustain this power is (222 MWth/200 MeV per fission) $R_{\text{fission}} = 7 \times 10^{18}$ fissions s^{-1} or

$$R_{\text{fisssion}}(\text{s}^{-1}) = 7 \times 10^{16} P_{\text{fission,el}}(\text{MW})$$

2. The number of fissions initiated per injected neutron $= \frac{k_{\text{eff}}}{(1-k_{\text{eff}})\nu}$. It follows that the neutron injection rate is given by

$$R_{\text{neutron-injection}}(\text{s}^{-1}) = \frac{7 \times 10^{16} P_{\text{fission,el}}(\text{MW})}{k_{\text{eff}}/[(1-k_{\text{eff}})\nu]}$$
$$= 7 \times 10^{16} P_{\text{fission,el}}(\text{MW}) \cdot \frac{(1-k_{\text{eff}})\nu}{k_{\text{eff}}}$$

$$(11.1)$$

For an electric power of 100 MW, $k_{\text{eff}} = 0.96$, and $\nu = 2.5$, $R_{\text{neutron-injection}} = 7.3 \times 10^{17}$ neutrons s^{-1}. The accelerator characteristics required to achieve this through proton spallation are then proton energy $= 1$ GeV, proton current $= 4$ mA (assuming the number of neutrons produced per 1 GeV proton is 30), proton power $= 4$ MW, electrical power $= 8.9$ MW (for an accelerator efficiency of 45%).

3. If these neutrons are produced by a d–t fusion source, each neutron results from a fusion reaction, so the fusion reaction rate is also equal to $R_{\text{neutron-injection}}$. Since the fusion neutrons have an energy of 14 MeV, they can be multiplied by a factor 2.5 in a Be blanket, such that the required fusion reaction rate can be reduced by this factor of 2.5, that is,

$$R_{\text{fusion}}(\text{s}^{-1}) = 2.8 \times 10^{16} P_{\text{fission,el}}(\text{MW}) \cdot \frac{(1 - k_{\text{eff}})\nu}{k_{\text{eff}}}$$

Since each fusion reaction generates an energy of 17.6 MeV, the fusion power is

$$P_{\text{fusion,th}} = 2.8 \times 10^{16} \cdot (17.6\,\text{MeV}) P_{\text{fission,el}}(\text{MW}) \cdot \frac{(1 - k_{\text{eff}})\nu}{k_{\text{eff}}}$$

or

$$P_{\text{fusion,th}}(\text{MW}) = 0.079 \cdot P_{\text{fission,el}}(\text{MW}) \cdot \frac{(1 - k_{\text{eff}})\nu}{k_{\text{eff}}}$$

4. The fusion gain is defined as $G_{\text{fusion}} = P_{\text{fusion,th}}/P_{\text{laser}}$ where P_{laser} is the input energy in the form of laser radiation. It follows that

$$P_{\text{laser}}(\text{MW}) = 0.079 \cdot \frac{1}{G_{\text{fusion}}} P_{\text{fission,el}}(\text{MW}) \cdot \frac{(1 - k_{\text{eff}})\nu}{k_{\text{eff}}}$$

If the conversion efficiency for converting electricity to light is $\varepsilon_{\text{laser}}$, then the electrical power required to generate the fusion power is

$$P_{\text{laser,el}}(\text{MW}) = 0.079 \cdot \frac{1}{\varepsilon_{\text{laser}} G_{\text{fusion}}} P_{\text{fission,el}}(\text{MW}) \cdot \frac{(1 - k_{\text{eff}})\nu}{k_{\text{eff}}} \quad (11.2)$$

For a fusion gain $G_{\text{fusion}} = 1$, and an electricity-to-light conversion efficiency of $\varepsilon_{\text{laser}} = 0.1$, the electrical power required to feed the laser is then $P_{\text{laser,el}} = 8.2\,\text{MW}$. The resulting fusion neutrons are used to drive a subcritical reactor with $k_{\text{eff}} = 0.96$ and produce 100 MW electrical energy. Equation (11.2) is the basic relation required to determine the laser power required to produce fusion neutrons, which are then used to drive the subcritical fission reactor. Conversely, for a fixed laser power and fission power, the efficiency can be expressed in terms of the fusion gain and k_{eff}, that is,

$$\varepsilon_{\text{laser}} = 0.079 \cdot \frac{1}{G_{\text{fusion}}} \cdot \frac{P_{\text{fission,el}}(\text{MW})}{P_{\text{laser,el}}(\text{MW})} \cdot \frac{(1 - k_{\text{eff}})\nu}{k_{\text{eff}}} . \quad (11.3)$$

This is essentially the relation plotted in Fig. 11.5 of the chapter. Keeping the ratio of the fission-to-laser power constant (=12.2), the above relation can be expressed in the form

$$\varepsilon_{\text{laser}} G_{\text{fusion}} M_{\text{fission}} \cong 1 , \quad (11.4)$$

where the multiplication factor M_{fission} of the subcritical system is given by $M_{\text{fission}} = k_{\text{eff}}/[(1 - k_{\text{eff}})\nu]$. Relation (11.4) combines the properties of the laser, the fusion and subcritical system together in a single formula. With $G_{\text{fusion}} = 1$ (scientific breakeven), $k_{\text{eff}} = 0.96$, $\nu = 2.5$, $\varepsilon_{\text{laser}} = 0.1$, the constant is 0.96, that is, approximately unity. If the fusion gain increases to 10 (engineering breakeven for $\varepsilon_{\text{laser}} = 0.10$), then k_{eff} can be reduced to $k_{\text{eff}} = 0.71$. If the fusion gain increases to 25, then k_{eff} can be reduced to $k_{\text{eff}} = 0.48$. This clearly demonstrates the greatly improved flexibility in subcritical systems for waste transmutation, when driven by fusion neutrons.

If the thermal power of a pure fusion system is 222 MW, and the laser (optical) power is 0.82 MW, then a gain of $G_{\text{fusion}} = 222/0.82 = 270$ is required to become economically practical.

High-Power Laser Production of PET Isotopes

L. Robson, P. McKenna, T. McCanny, K.W.D. Ledingham, J.M. Gillies, and
J. Zweit

Department of Physics, University of Strathclyde, Glasgow, G4 0NG, Scotland, UK
l.robson@phys.strath.ac.uk
p.mckenna@phys.strath.ac.uk

12.1 Introduction

Recent experiments have demonstrated that laser–solid interactions at intensities greater than 10^{19} W/cm^2 can produce fast electron beams of several hundred MeV [1], tens of MeV γ-rays [2, 3], up to 58 MeV proton beams [4, 5], and heavier ions [6] of up to 7 MeV/nucleon. One of the potential applications of the high-energy proton beams is the production of radioactive isotopes for positron emission tomography (PET). PET is a form of medical imaging requiring the production of short-lived positron emitting isotopes ^{11}C, ^{13}N, ^{15}O, and ^{18}F, by proton irradiation of natural/enriched targets using cyclotrons. PET development has been limited because of the size and shielding requirements of the nuclear installations. Recent results have shown when an intense laser beam interacts with solid targets, tens of MeV protons capable of producing PET isotopes are generated [7, 8, 9].

In the following section, the principles of the PET technique are introduced, including the key applications and current methods of producing PET isotopes. High-power laser–plasma interactions leading to the production of multi-MeV protons capable of producing PET isotopes are discussed in detail, including the physics of the laser–plasma ion acceleration and recent results using the VULCAN petawatt laser at the Rutherford Appleton Laboratory to produce PET isotopes ^{11}C and ^{18}F. In this experiment, for the first time laser production of ^{18}F by a (p,n) reaction on ^{18}O and the subsequent synthesis of 2-[^{18}F] was reported by Ledingham et al. [8] and the details are presented here.

In the final section of this chapter, the potential for developing on-site, easy-to-shield, compact laser technology for this purpose will also be discussed, describing two proposed laser systems that may be able to produce PET isotopes on a scale similar to cyclotrons.

L. Robson et al.: *High Power Laser Production of PET Isotopes*, Lect. Notes Phys. **694**, 191–203 (2006)
www.springerlink.com   © Springer-Verlag Berlin Heidelberg and European Communities 2006

12.2 Positron Emission Tomography

PET is a powerful medical diagnostic and imaging technique requiring the production of short-lived (2 min – 2 hour) posistron emitting isotopes. The PET process involves the patient receiving an injection of a pharmaceutical labeled with a short-lived $\beta+$ emitting source which collects in areas of high metabolic activity within the body such as tumors. Thus, specific sites in the body can be imaged by detecting the back-to-back 511 keV positron–electron annihilation γ-rays emitted from the radiopharmaceutical. Some of the key applications of PET are imaging/diagnosing blood flow, amino acid transport, and tumors. The principal tracers used in the PET technique are ^{11}C, ^{13}N, ^{15}O, and ^{18}F. Many chemical compounds can be labeled with positron emitting isotopes and their biodistribution can be determined by PET imaging as a function of time. The most commonly used radiopharmaceutical is 2-fluoro-2-deoxyglucose 2-[^{18}F]FDG. Various biochemical events including glucose metabolism can be directly assessed in patients to reveal changes in the metabolic activity resulting from disease progression and therapeutic intervention. Over the last few years the value of PET FDG in the management of cancer patients has been widely demonstrated. Figure 12.1 highlights the success rate of PET in diagnosing lung cancer compared with conventional X-ray computed tomography (CT) scanning.

PET isotopes are produced using energetic proton beams produced by cyclotrons [10, 11] or van de Graafs via (p,n) or (p,α) reactions. Table 12.1 lists the reactions used to create the isotopes, the associated reaction threshold, the half-life of the product, and the peak cross sections. Proton-induced reactions are favored since the product is a different chemical element to the target (and therefore can be easily separated by chemistry). Thus, after subsequent synthesis of the radioisotope, the patient can be injected with the minimum amount of foreign material. The separation of the isotope is described in detail later.

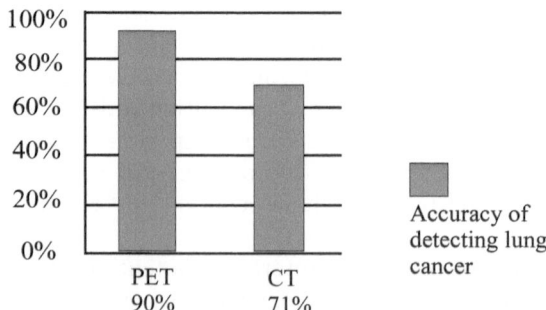

Fig. 12.1. Accuracy of PET in detecting lung cancer compared with X-ray CT scanning [12]

Table 12.1. PET radioisotopes production reactions

Nuclear Reaction	Half-life	Q(MeV)	Peak Cross Section (mb)	Radiation Measured
^{15}N(p,n)^{15}O	9.96 min	3,53	200	$\beta + 100\%$
^{16}O(p,α)^{13}N	123 s	5.22	140	$\beta + 100\%$
^{14}N(p,α)^{11}C	20.34 min	2.92	250	$\beta + 99\%$
^{11}B(p,n)^{11}C	20.34 min	2.76	430	$\beta + 99\%$
^{18}O(p,n)^{18}F	109.7 min	2.44	700	$\beta + 97\%$

One of the main factors limiting the wider use of FDG PET imaging is the requirement for expensive infrastructure at the heart of which lies the cyclotron and the associated extensive radiation shielding. A more simplified approach to isotope production would be to develop a miniaturized, on-site resource with eventual capability similar to that of a cyclotron. As was stated previously, recent results show when an intense laser beam ($I > 10^{19}$ W/cm^2) interacts with solid targets, beams of MeV protons capable of producing PET isotopes are generated. Recent reports have concentrated on some preliminary work carried out by this group [7, 8] and Fritzler et al. [9] on the production of PET isotopes using a high-power laser.

As early as the seventies, it was proposed [13] that laser-driven electron acceleration was possible using intense laser light to produce a wake of oscillations in a plasma. Recently, 200 MeV electrons were measured using a compact high repetition rate laser [1], while electrons with an energy up to 350 MeV have been reported using the VULCAN petawatt laser [14]. In addition, the generation of monoenergetic electron beams from intense laser–plasma interactions has also been reported [15, 16, 17]. laser–plasma–based accelerators can deliver accelerating gradients more than 1,000 times higher than conventional accelerator technology, and on a compact scale. This increase in accelerating gradient is the key factor to reducing the scale and therefore the associated cost over current conventional accelerators.

After an extended program of research, Ledingham et al. [8] reported on the generation of intense short-lived PET sources, ^{11}C and ^{18}F, using the VULCAN petawatt laser beam and also, for the first time, the synthesis of 2-[^{18}F]-fluorodeoxy glucose, the "workhorse" of PET technology from laser-driven ^{18}F using an enriched water target. Here we review the details of the experimental procedure and findings, as well as introducing the principles of laser–plasma proton production.

12.3 Proton Acceleration with a High-Intensity Laser

Recent advances in laser technology with the introduction of chirped pulse amplification [18] (CPA) have led to the development of multiterawatt pulsed laser systems in many laboratories worldwide. In CPA, a laser pulse of the order of femtoseconds or picoseconds is temporally stretched by three to four orders of magnitude using dispersive gratings, thus preventing damage to the laser amplifying medium from nonlinear processes at high intensities. After amplification, these laser pulses are recompressed to deliver 10^{18-20} W/cm^2 on target. Proposed techniques, including optical parametric chirped pulse amplification (OPCPA) [19, 20], promise to extend the boundaries of laser science into the future and also reduce the large lasers used presently to compact tabletop varieties. Future developments of specific laser systems are discussed in detail later.

High-intensity laser radiation may now be applied in many traditional areas of nuclear science. As the laser intensity and associated electric field is increased, then the electron quiver energy, the energy a free electron has in the laser field, increases dramatically. Thus, when laser radiation is focused onto solid and gaseous targets at intensities $> 10^{18}$ W/cm^2, electrons quiver with energies greater than their rest mass (0.511 eV) creating relativistic plasmas [21]. At these intensities, the Lorentz force $-e(\boldsymbol{v} \times \boldsymbol{B})$ due to the laser interacting with charged particles produces a pondermotive force allowing electrons to be accelerated into the target in the direction of laser propagation. The resulting electron energy distribution can be described by a quasi-Maxwellian distribution yielding temperatures (kT) of a few MeV [22].

The mechanism responsible for laser–plasma ion acceleration is currently the subject of intense research in many laboratories worldwide. The protons emanate from water and from hydrocarbons as contaminant layers on the surfaces of the solid targets. These contamination layers are due to the poor vacuum (approximately 10^{-5} Torr) achievable in the target chambers in which these experiments are normally carried out. The main mechanism thought to be responsible for proton acceleration is the production of electrostatic fields due to the separation of the electrons from the plasma ions. Proton beams are observed both in front of (blow-off direction) and behind (straight-through direction) the primary target. In front of the target, ion beams are observed from the expansion of the plasma generated on the target surface, produced either by a prepulse or by the rising edge of the main pulse itself, also known as "blow-off" plasma, directed normally to the target surface.

Several acceleration mechanisms have been proposed to describe where the protons in the straight-through direction originate, the front surface, back surface, or both. One such mechanism is the Target Normal Sheath Acceleration (TNSA) [23]. In this scheme, shown in Fig. 12.2, the ion acceleration mechanism results from the cloud of hot electrons (generated in the blow-off plasma from the laser prepulse interacting with the front surface of the target) traveling through the target and ionizing the contaminant hydrogen layer

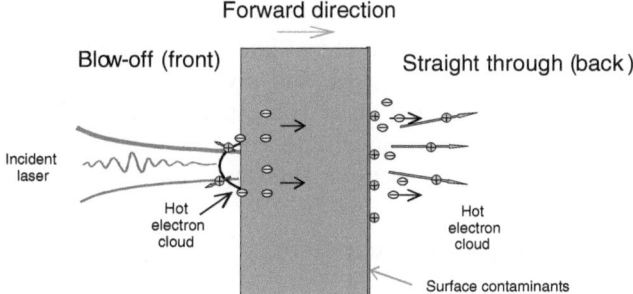

Fig. 12.2. Pictorial representation of the Target Normal Sheath Acceleration (TNSA) [23] scheme. Protons are accelerated in the blow-off direction from the thermal expansion of the hot plasma, and electrons are pondermotively driven into the target, creating an electrostatic sheath on the rear side, leading to acceleration of the proton contaminants from the back surface

on the back surface of the target. The protons are then pulled off the back surface by the cloud of electrons and accelerated normally to the target to tens of MeV's on the order of micrometers. It has also been shown that the accelerating gradient is dependent on the plasma scale length [23]. The initial laser pre-pulse may be of the order of 10^{-6} ($I \sim 10^{12-14}\,\mathrm{W/cm^2}$) of the main laser pulse, sufficient to ionize the front surface of the target. Thus, at the back of a suitable thickness target where no preplasma is formed, the accelerating field is greater, resulting in higher energy ions. Recent studies have reported on direct experimental evidence of back-surface ion acceleration from laser-irradiated foils by using sputtering techniques to remove contaminants from both the front and back surfaces [24]. It has also been proposed that the protons are accelerated via an electrostatic sheath formed on the front surface of the target and dragged through the target to produce a proton beam at the rear of the target [4]. Comparative reports on the ion acceleration schemes can be found in [25, 26].

Proton energies with an exponential distribution up to 58 MeV have been observed [5] for a laser pulse intensity of $3 \times 10^{20}\,\mathrm{W/cm^2}$ and production of greater than 10^{13} protons per pulse has been reported [27]. With the VULCAN laser at the Rutherford Appleton Laboratory (RAL) delivering petawatt powers, it is now possible to demonstrate the potential for high-power lasers to produce intense radioactive sources.

12.4 Experimental Setup

The petawatt arm of the VULCAN Nd:Glass laser at RAL was employed in the experimental study by Ledingham et al. [8]. The 60-cm beam was focussed to approximately 5.5-μm-diameter spot using a 1.8-m focal length off-axis parabolic mirror, in a vacuum chamber evacuated to approximately

10^{-4} mbar. The energy on target was between 220 and 300 J while the average pulse duration was approximately 1 ps. The peak intensity was of the order of 2×10^{20} W/cm^2. Aluminum, gold, and mylar foil targets of various thicknesses (1–500 μm) were irradiated by the p-polarized laser beam incident at an angle of 45°. As stated previously, the protons emanated from water and hydrocarbon contamination layers on the target surfaces.

12.4.1 Proton Energy Measurements

To measure the energy spectra of the accelerated protons, nuclear activation techniques were employed. Copper stacks (5 cm × 5 cm) were positioned along the target normal direction and exposed to the protons accelerated from both the front and back surfaces of the primary target foil. Figure 12.3 shows an image of the experimental setup inside the chamber.

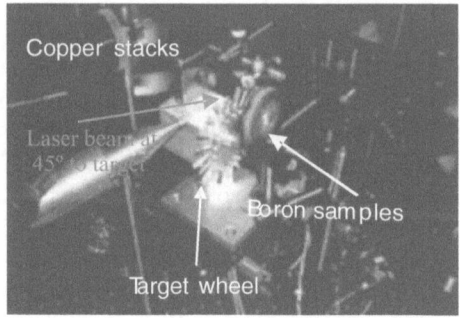

Fig. 12.3. Image of inside the target chamber showing the incident laser beam directed onto varying thicknesses of materials of foils held in a target wheel. The Copper stacks for proton energy measurements and Boron samples for ^{11}C production are shown

The activity in the copper foils from the ^{63}Cu(p,n)^{63}Zn reaction with a half-life of 38 min was measured in a 3″ × 3″ NaI coincidence system setup to detect the annihilation photons at 511 keV. The efficiency of the system was measured using a calibrated ^{22}Na source; thus, the absolute activity; that is, the number of (p,n) reactions in each copper piece could be determined. The measured activity in the foils from the ^{63}Cu(p,n)^{63}Zn, convoluted with the reaction cross section (shown in Fig. 12.4a) and proton stopping powers was used to produce the energy distributions shown in Fig. 12.4b.

As the laser intensity increases, protons are produced with much higher energies and, in addition, proton-induced reactions with higher Q-values, such as (p,2n), (p,3n), and (p,p+n) reactions can be produced. Measurement of these proton-induced reactions in a single layer of copper foil has been identified [29] as an alternative method to the use of copper stacks in diagnosing accelerated proton energy spectra from laser–plasma interactions and allows

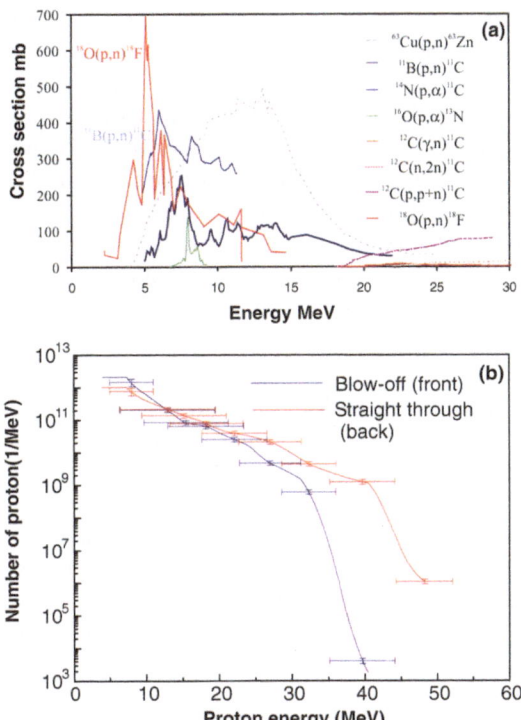

Fig. 12.4. (a) Experimentally measured cross sections [28] for the nuclear reactions used to diagnose the proton spectra. Also shown are the cross sections for the nuclear reactions described for the production of PET isotopes. (b) Typical proton spectra in front of and behind a 10-μm Al target. The spectra were quasi exponential with the highest energy protons measured behind the target. The number of protons generated per laser shot at about 300 J and 2×10^{20} W/cm^2 was typically 10^{12}

the proton beam to be measured simultaneously with other experiments. An example of this is shown in Fig. 12.3 above, where the boron activation sample for ^{11}C production is covered by a thin piece of copper.

It is evident from Fig. 12.4b that more energetic protons are observed in the straight-through (back) direction, having energies up to approximately 50 MeV, whereas the maximum proton energy in the blow-off (front) direction is approximately 40 MeV. In addition, it is important to point out that the end point of the proton spectrum (back) has increased from about 30 MeV for a 100 TW [7] laser (based on previous experiments carried out by this group) to about 50 MeV for the close to petawatt laser.

12.4.2 ^{18}F and ^{11}C Generation

The isotope ^{18}F was generated from a (p,n) reaction on ^{18}O enriched (96.5%) target. The enriched ^{18}O targets were irradiated in the form of 1.5 mL of

[^{18}O]H$_2$O placed in a 20-mm-diameter stainless steel target holder. The holder was assembled with a 100-μm aluminum window, and secured with a stainless steel clamping plate. For the production of ^{11}C, the copper stacks described above were replaced by boron samples (5 cm in diameter and 3 mm thick). After irradiation, the boron targets were removed from the vacuum chamber and the ^{11}C activity produced by the (p,n) reaction on ^{11}B was measured in the coincidence system up to 2 hours after the laser shot, a safety precaution because of the high activity. The counting rate was determined at time zero and converted to Bq using a calibrated ^{22}Na source.

12.4.3 Target Selection

In order to determine the thickness of primary target that generated the highest activity sources, the ^{11}C activity generated in the secondary ^{11}B targets was measured as a function of sample material and thickness. The ratio of the back-to-front activities is shown in Fig. 12.5. This was carried out using the production of the PET isotope ^{11}C rather than the more novel ^{18}F because of the cost of carrying out systematic work using the very expensive separated ^{18}O isotope as a target. It is clear from Fig. 12.5 that very thin targets provide the highest activity sources when the total activity produced per laser shot is the sum of the back and front activities.

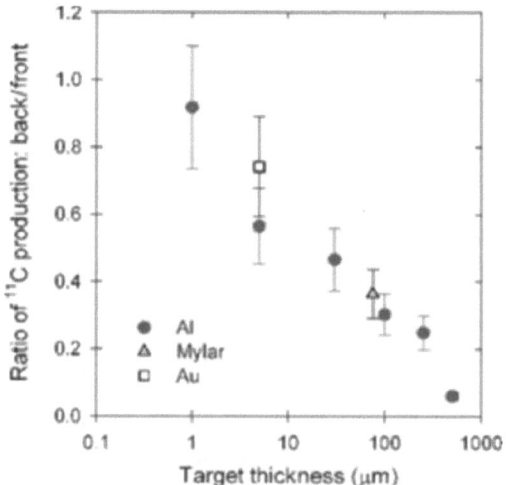

Fig. 12.5. Back/front ratio of ^{11}C from the (p,n) reactions on ^{11}B as a function of target thickness. At the highest pulse energy on target 300 J the ^{11}C activity maximally was about 6×10^6 Bq per shot on each side. This is greater than 10^7 Bq in total. Behind the target the ^{11}C activity decreased with increasing thickness while for similar energy on target the activity in front of the target was largely independent of target thickness or indeed of material

12.5 Experimental Results

12.5.1 ^{18}F and ^{11}C Production

It was reported earlier that ^{18}F is the most widely used tracer in clinical PET today because of its longer half-life allowing for the synthesis of a number of samples within a half-life decay of the isotope and because fluorine chemistry is readily introduced in many organic and bioinorganic compounds. Therefore, it was important that Ledingham et al. [8] determined how much ^{18}F could be produced per laser shot. The isotope was generated from a (p,n) reaction on ^{18}O-enriched (96.5%) target. At the highest laser pulse energies (300 J), 10^5 Bq total activity of ^{18}F was produced (shown later).

The measured half-life for the ^{18}F source is shown in Fig. 12.6. The half-life of 110 ± 3 min was determined over more than three half-lifes and demonstrates the purity of the ^{18}F source generated and agrees closely with the generally accepted value (109 min). The measured half-life of ^{11}C was approximately 20 min and is not shown but has been discussed in detail previously [7] and agrees well with the accepted value (20.3 min).

12.5.2 Automated FDG Synthesis

For the first time the synthesis of 2-[^{18}F]FDG using laser-induced ^{18}F activity was demonstrated by Ledingham et al. [8]. The synthesis was based on the method developed by Hamacher et al. [30], shown in Fig. 12.7 adapted for the production of ^{18}F from a remote cyclotron source and the FDG Coincidence-Kit Based Synthesizer (GE Medical Systems). Briefly, a mannose triflate precursor was fluorinated following the recovery of H$_2^{18}$O. Subsequent hydrolysis and column purifications yield 2-[^{18}F]FDG. The radiochemical purity

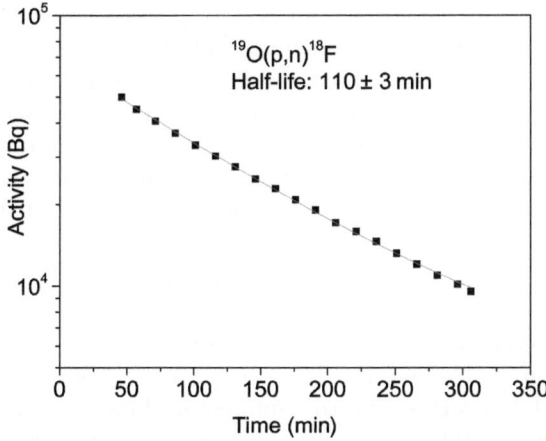

Fig. 12.6. The measured half-life for ^{18}F. The value was close to the accepted one indicating the purity of the source produced

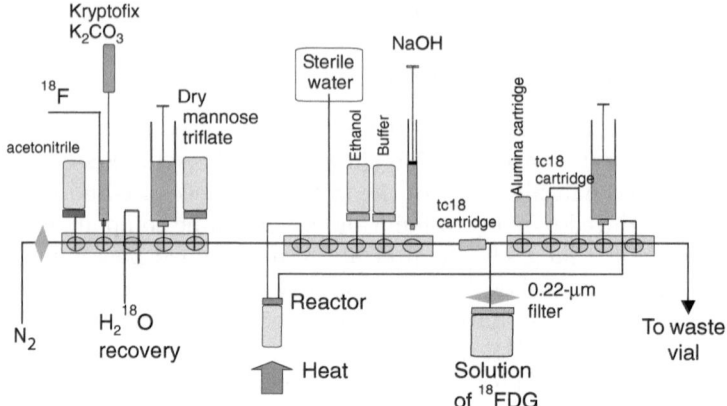

Fig. 12.7. FDG synthesis process

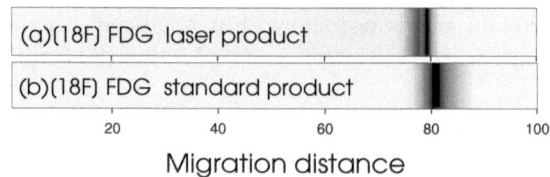

Fig. 12.8. Radio thin layer chromatography analysis showing the laser-produced [^{18}F] FDG and standard cyclotron [^{18}F] FDG. The analysis shows that both sources are radiochemically pure

and yields were determined by quantitative radio thin-layer chromatography (TLC) and are displayed in Fig. 12.8. The laser- and cyclotron-induced markers on the TLC trace represents a 1-μL sample of 2-[^{18}F]FDG taken from each product for both the laser- and cyclotron-induced activities. The analysis was carried out using an Instant Imager electronic autoradiography system (Packard, USA). In Fig. 12.8, both products (radio thin-layer chromatogram of 2-[^{18}F]FDG synthesized from (a) the laser-induced F-18 activity and (b) the cyclotron-induced F-18 activity) show the same migration distance of approximately 75 mm with an R_f (retention factor) value of 0.63. The TLC analysis shows that both sources are radiochemically pure and similar to each other.

12.5.3 Activity of Laser-Produced PET Sources

Figure 12.9 summarizes the measurements to date in this program of research into laser-driven ^{11}C and ^{18}F PET isotope production on VULCAN. The circular points [^{11}C] correspond to a number of different laser irradiances and pulse energies up to 300 J with a pulse duration approximately 1 ps. The single triangular point is the activity from the ^{18}F measurements at the highest laser pulse energy. The shaded areas at the top of the graph provide an indication

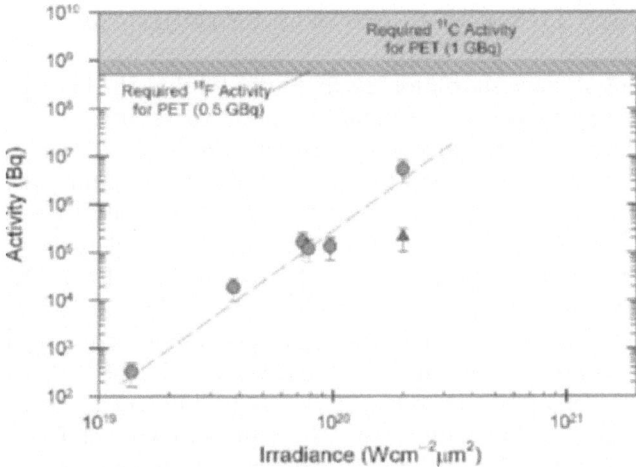

Fig. 12.9. The total activity (front and back) generated by a single laser shot for both ^{11}C and ^{18}F as a function of laser irradiance with pulse energies from 15 to 300 J. The circles refer to ^{11}C production and the single triangular point for ^{18}F was measured at the highest energy. The activity measured for ^{18}F is lower than that of ^{11}C since the amount of initial activation sample ($[^{18}$O$]$H$_2$O) was much less in the ^{18}F case and extra thick protective windows [50 μCu] were placed over the water targets to prevent any possible rupture under vacuum which reduced the proton energy and hence any (p,n) activity

for the level of required ^{18}F activity (0.5 GBq) from which an ^{18}F-FDG patient dose would be generated and the required ^{11}C activity (1 GBq), for example, in the form of $[^{11}$C$]$CO. However, it is important to note that in order to synthesize a ^{11}C-labeled compound for a patient study, the total amount of ^{11}C required would be in the 1–3 GBq range, depending on the yields and the duration of the synthesis process.

It can be seen that if the fit to the experimental data is extrapolated on this graph then at 10^{21} Wcm^{-2} μm^2 (equivalent to a total pulse energy of 1 kJ), sufficient activity of ^{11}C in one laser shot may be generated, equivalent to the minimum required patient dose for PET sources. It should be pointed out that experiments have yet to be conducted at these higher laser irradiances and this statement of extrapolation is made with caution. In addition, the discussion refers only to the pulse and focal conditions of the VULCAN laser and other lasers with different pulse and contrast ratio parameters may behave differently. The final section of this chapter addresses the issue of increasing the isotope activity.

12.6 Future Developments and Conclusions

Ledingham et al. have shown for the first time [8] that the laser-produced PET isotope ^{18}F can undergo successful synthesis to produce the radiopharmaceutical 2-fluoro-2-deoxyglucose 2-[^{18}F]FDG. However, as was shown earlier, the activity of the sources was below the required minimum level for patient doses. Although the results reviewed here were obtained from a large single shot laser, it is important to highlight the progress made using compact high repetition rate lasers. Fritzler et al. [9] have calculated that 13 MBq of ^{11}C can be generated using the LOA "tabletop" laser (1 J, 40 fs) 6×10^{19} W/cm^2 after 30 min at 10 Hz and that this can be extended to GBq using similar lasers with kHz repetition rates. Alternatively, at JanUSP (Livermore) using a single pulse (8.5 J, 100 fs, 800 nm) at 2×10^{20} W/cm^2, 4.4 kBq of ^{11}C was generated from a single laser shot [31]. Using a compact laser with similar specifications at 100 Hz after 30 min, this would amount to close to GBq. A compact "tabletop" laser system has recently been designed by Collier and Ross [32]. This OPCPA system is envisaged to be capable of delivering 6 J in 50 fs at 100 Hz at optical irradiances between 10^{20} and 10^{21} W/cm^2-μm^2. OPCPA technology is at an early stage of development but sufficient progress has been made to make one reasonably optimistic that the above specification is attainable. Such a laser would be capable of producing GBq activities of PET isotopes in 30 min. In addition, the small scale POLARIS [33] all diode pumped petawatt laser currently being built at the Friedrich-Schiller University of Jena has the potential to deliver 10^{21} W/cm^2 ($\tau = 150$ fs, $E = 150$ J, $\lambda \sim 1\,\mu$m) with a repetition rate of 0.1 Hz.

In conclusion, it has been shown that very intense PET sources of ^{11}C and ^{18}F are produced using a large petawatt laser and also, for the first time, the synthesis of 2-[^{18}F]-fluorodeoxy glucose, the "workhorse" of PET technology. The potential for developing the technology for on-site, easy-to-shield compact laser technology has also been discussed. In addition to increasing the laser intensity on target and developing tabletop laser systems at high repetition rate to integrate over many shots, other parameters have been determined that should increase the activity of the laser-produced sources. Very recently, Nakamura et al. [34] reported that when a polymer-coated metal target was irradiated with laser pulses of 10^{17} W/cm^2, a significant enhancement ($\times 80$) of fast protons was produced over the uncoated target and hence there is every likelihood that significant increase of proton production and hence PET isotope activity will be produced when the layers of contaminants are replaced by controlled surfaces of hydrogen atoms.

Acknowledgments

The authors would like to acknowledge the invaluable contribution of colleagues at the Central Laser Facility – Rutherford Appleton Laboratory, the

Christie Hospital, UMIST, the University of Glasgow, Imperial College, and the Queen's University. PMcK gratefully acknowledges the award of a Royal Society of Edinburgh Personal Fellowship.

References

1. V. Malka et al.: Science **298**, 1596 (2002)
2. M.I.K. Santala et al.: Phys. Rev. Lett. **84**, 1459 (2000)
3. B. Liesfeld et al.: Appl. Phys. B **79**, 1047 (2004)
4. E.L. Clark et al.: Phys. Rev. Lett. **84**, 670 (2000)
5. R.A. Snavely et al.: Phys. Rev. Lett. **85**, 2945 (2000)
6. P. McKenna et al.: Phys. Rev. Lett. **91**, (2003)
7. I. Spencer et al.: Nucl. Instr. Meth. **183**, 449 (2003)
8. K.W.D. Ledingham et al.: J. Phys. D: Appl. Phys. **37**, 2341 (2004)
9. S. Fritzler et al.: Appl. Phys. Lett. **83**, 3039 (2003)
10. K. Kettern et al.: Appl. Rad. Isot. **60**, 939 (2004)
11. www.manpet.man.ac.uk, 2004,
12. http://www.nuc.ucla.edu/pet/
13. T. Tajima and J.M. Dawson: Phys. Rev. Lett. **43**, 267 (1979)
14. S.P.D. Mangles et al.: Phys. Rev. Lett. (submitted) 2004
15. S.P.D. Mangles et al.: Nature, **431**, 535 2004
16. J. Faure et al.: Nature, **431**, 541 (2004)
17. C.G.R. Geddes et al.: Nature, **431**, 538 (2004)
18. D. Strickland and G. Mourou: Opt. Commun. **56**, 219 (1985)
19. I.N. Ross: Laser Part Beams **17**, 331 (1999)
20. A. Dubietis et al.: Opt. Commun. **88**, 437 (1992)
21. D. Umstadter: Phys. Plasmas **8**, 1774 (2001)
22. I. Spencer et al.: Rev. Sci. Inst. **73**, 3801 (2002)
23. S.C. Wilks et al.: Phys. Plasmas. **8**, 542 (2001)
24. M. Allen: PhD thesis 2004 Laser Ion Acceleration From the Interaction of Ultra-Intense Laser Pulse With Thin Foils: LLNL - UCRL-TH-203170
25. D. Umstadter: J. Phys. D: Appl. Phys. **36**, R151 (2003)
26. M. Zepf et al.: Phys. Plasmas. **8**, 2323 (2001)
27. S.P. Hatchett ct al.: (2000), Phys. Plasmas. **7**, 2076
28. IAEAND.IAEA.OR.AT/exfor: (2004), EXFOR Nuclear Reaction Database
29. J.M. Yang et al.: Appl. Phys. Lett. **84**, 675 (2004)
30. K. Hamacher et al.: J. Nucl. Med. **27** (1986)
31. P.K. Patel: private communication
32. J.L. Collier and Ross IN: private communication
33. www.physik.uni-jena.de/qe/Forschung/F-Englisch/Petawatt/Eng-FP-Petawatt.html
34. K.G. Nakamura et al.: Conference Proceedings-Field Ignition High Field Physics (Kyoto, Japan), p. 2724 (2004)

Part IV

Nuclear Science

13

Nuclear Physics with High-Intensity Lasers

F. Hannachi, M.M. Aléonard, G. Claverie, M. Gerbaux, F. Gobet, G. Malka, J.N. Scheurer, and M. Tarisien

Centre d'Etudes Nucléaires de Bordeaux Gradignan, Université Bordeaux 1, le Haut Vigneau, 33175 Gradignan cedex, France
hannachi@cenbg.in2p3.fr

13.1 Introduction

A laser is a unique tool to produce plasma and photon beams or particle beams with very high fluxes and very short durations. Both aspects are of interest for fundamental nuclear physics studies.

In plasma the electron–ions collisions are predicted to modify atomic level widths and possibly nuclear level ones. This is of prime importance for the population of isomeric states and the issue of energy exchange between the nucleus and the electronic system. Several nuclear observables such as lifetimes and beta decay probabilities are sensitive to the charge state and to the excitation energy. Furthermore, with a laser it is possible to produce electric and magnetic fields strong enough to change the binding energies of electronic states. If nuclear states happen to decay via internal conversion (IC) through these perturbed states, a modification of their lifetimes will be seen. In this chapter we will report on the status of ongoing experiments on these different topics in our group at CENBG: The search for the NEET process in ^{235}U, the population of the ^{181}Ta (6.2 keV, 6.8 μs) isomeric state in plasma, and the influence of the laser electric field on nuclear properties. These experiments illustrate some of the new opportunities offered by the available laser facilities to extend investigations of nuclear properties under exotic conditions.

13.2 Search for NEET in ^{235}U

The excitation of nuclear levels by the transfer of energy from the atomic part to the nuclear part of an atom is the subject of a large number of investigations. Their goal is to find an efficient mechanism to populate nuclear isomers in view of further applications to energy storage and development of lasers based on nuclear transitions. This process called NEET (Nuclear Excitation by Electronic Transition) has been first suggested by Morita for the excitation of a level at approximately 13 keV in ^{235}U [1]. The NEET process is

F. Hannachi et al.: *Nuclear Physics with High-Intensity Lasers*, Lect. Notes Phys. **694**, 207–216 (2006)
www.springerlink.com   © Springer-Verlag Berlin Heidelberg and European Communities 2006

the inverse of the resonant nuclear internal conversion between bound atomic states (BIC) which has been demonstrated in ^{125}Te [2]. In this case, the well-known variation of the electron-binding energy with Q, the charge state of an ion, was used to reach the condition $Q = 45^+$ in which the energy matching between the atomic energy transition and the nuclear energy is realized. In the search for NEET in a plasma induced by a laser, the laser beam is used to create a dense and hot plasma of uranium matter. In the plasma, U atoms are ionized and have a charge state distribution that depends on the plasma temperature. Furthermore, to each of the charge states correspond several different configurations due to the coupling between the electron spins. Each atomic configuration corresponds to a particular set of atomic energy transitions. Some configurations have transitions that match more or less closely the nuclear energy transition between the ground state and the first excited state in ^{235}U (see Fig. 13.1). The nucleus can absorb only the virtual photon emitted from the atomic transition if the energy mismatch between the two energy transitions is of the order of the width of the system. Because of electron–ion collisions in the hot plasma, the widths of the excited atomic levels are strongly increased. As an example, the natural width of a $5d$ hole in the U atom is of the order of 10^{-5} eV. In a plasma at a temperature of 100 eV, this width becomes dominated by the Stark broadening effect and reaches values as large as 20 meV. This greatly enhances the possible matching between the atomic and the nuclear energy transitions.

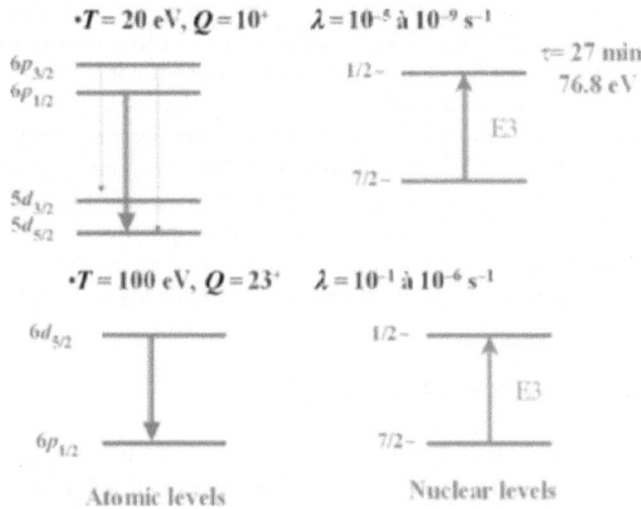

Fig. 13.1. Energy levels in ^{235}U at two different plasma temperatures T where electronic transition energies are nearly resonant with the nuclear transition energy (± 4 eV). The corresponding predicted nuclear excitation rates λ are given (from reference [3])

In ^{235}U the nuclear transition between the ground state and the first excited state is an E3 transition whose characteristics are known with poor accuracy. The transition energy is $E_n = (76.8 \pm 0.8)$ eV. The half-life of the $1/2^-$ excited level is 26.8 min depending slightly on the chemical state of U. The internal conversion coefficient is very large approximately 10^{20} but has never been measured. In spite of these uncertainties it has been shown that the atomic transitions $6p - 5d$ in U ions with $Q = 10$ and $6d_{5/2} - 6p_{1/2}$ in U ions with $Q = 23$ have energies nearly equal with E_n. Owing for combined uncertainties on E_n and on the calculated atomic energies leading to an energy mismatch of 4 eV, theoretical values of the nuclear excitation rate range between 10^{-9} and 10^{-5} s^{-1} for the first group of transitions and between 10^{-5} and 10^{-1} s^{-1} for the second one [3]. The calculations have been performed for an electronic density of 10^{19} cm^{-3} in the plasma.

Several experiments have been done over the last 30 years to observe the excitation of the extremely low energy level at 76 eV in ^{235}U using a pulsed high-intensity laser beam. In these experiments, the plasma generated in the interaction of the laser with the U target was collected on a catcher foil subsequently placed in front of an electron multiplier. The excitation of the isomeric level was detected by means of the internal conversion electrons from its decay. The results of these experiments are contrasted. Using a 1 J CO_2 laser and a target of natural uranium, Yzawa et al. [4] have observed a strong signal of delayed low-energy electrons, attributed to the decay of the 76 eV level after excitation by a NEET process. Goldansky and Namiot [5] pointed out that in these experimental conditions the excitation probability by NEET should be very small and they proposed a new interpretation of the excitation of the nucleus in terms of NEEC. NEEC is the excitation of the nucleus by the capture of a free electron into a bound orbital. The energy gained by the system can be resonantly transferred to the nucleus. Arutynyan et al. [6] attempted a similar experiment with a CO_2 laser (5 J, 200 ns) and a ceramic target 6% enriched in ^{235}U but failed to observe the excitation of the isomeric state. In a second experiment the same group observed an excitation of the isomeric state detected in a plasma induced by a high-intensity beam of 500 keV electrons. The temperature of the plasma was of the order of 20 eV. It was shown in [3] that this positive result was most probably due to a direct excitation of the U nuclei by inelastic electron scattering from the incident beam rather than to a NEET or a NEEC mechanism.

In fact, it is very difficult to compare these results because of the lack of details on the experimental conditions such as laser beam focusing, the plasma temperature, number of collected atoms on the catcher foil, electron detection efficiency, and parasitic electron emission phenomena. Finally, we mention a more recent attempt to excite the isomeric level in ^{235}U by Bound and Dyer [7]. In this case, a CO_2 laser interacting with a 93% enriched U target generated a U vapor which was illuminated by a ps laser to create a plasma. The intensity of the ps laser beam was deduced from a measurement of the charge state distribution of the U ions in the plasma. They found an intensity ranging between 10^{13} and 10^{15} W/cm^2. The U ions were deposited on a plate and the delayed electron emission was analyzed. An electric field permitted the separation between neutral and ionized U species. A time of flight measurement of the collected U ions provided with a crude estimation of the number of ions in charge states ranging between 1$^+$ and 5$^+$. In spite of the very careful experimental method, no excitation of the isomeric level was observed in the experiment. We have performed at CENBG an experiment to search for the nuclear excitation of the 76 eV isomeric level in ^{235}U in a plasma induced by a 1 J, 5 ns, Nd Yag laser, at a wavelength of 1.06 μm. Targets 93% enriched in ^{235}U were used. Eighty percent of the laser energy was found in a 40-μm-diameter spot leading to a laser intensity, $I \sim 10^{13}$ W/cm^2. A sketch of the experimental setup is shown in Fig. 13.2.

In a first step, a ^{235}U plasma is formed by interaction of the laser beam with a U target. The plasma expands in vacuum, and the U atoms are collected on a catcher foil. In a second step the catcher foil is quickly moved in front of an electron detector to search for the decay by internal conversion of the level at 76 eV in ^{235}U, possibly excited in the plasma by a NEET process. The nuclear excitation is signed by the detection of electrons with a maximum energy of 57 eV. The number of electrons emitted versus the time elapsed after the end of the irradiation must follow an exponential law with a half-life period equal to 26.8 min.

In this experiment, the improvement in the sensitivity to the IC electrons emitted in the decay of the isomer is due to the reduction of the signal associated with the alpha radioactivity of U isotopes and with the careful study and subtraction of the exoelectron emission from the collector (see [8] for details).

We have set a lower limit on the excitation rate of the nuclear isomeric state in a plasma induced by a laser focused at an intensity of 10^{13} W/cm^2. The minimum number that could have been detected has been determined in the following way. We have added to the curve in Fig. 13.3a theoretical distribution of electrons $m_e(t)$ according to the decay of an isomeric state with a half-life of 26.8 min:

$$m_e(t) = m_0 \exp - ((0.5 \ln 2)t/26.8), \tag{13.1}$$

where t is expressed in minutes and m_0 is the number of electrons detected during the first 30 s after the last laser pulse. The number m_0, was decreased step by step. A fit of the distribution shown in Fig. 13.3b was used to obtain the minimum number m_0, which can be extracted from the data at a 1σ level of error. We find $m_0 = 10 \pm 3$ and a half-life 24 ± 4 min.

From the measurement of the rate of alpha particles detected in the Si detectors, we can deduce the number of ^{235}U atoms that have been collected [$N = (4.6 \pm 0.1)10^{17}$ atoms] and set a limit on the rate of nuclear excitation. We find $\lambda = 5.910^{-6}$ s^{-1}.

In summary, we have searched for the excitation in a plasma induced by laser of the isomeric level at 76 eV in ^{235}U. At a laser intensity of 10^{13} W/cm^2, an upper limit has been set on the nuclear excitation rate $\lambda = 5.9 \times 10^{-6}$ s^{-1}.

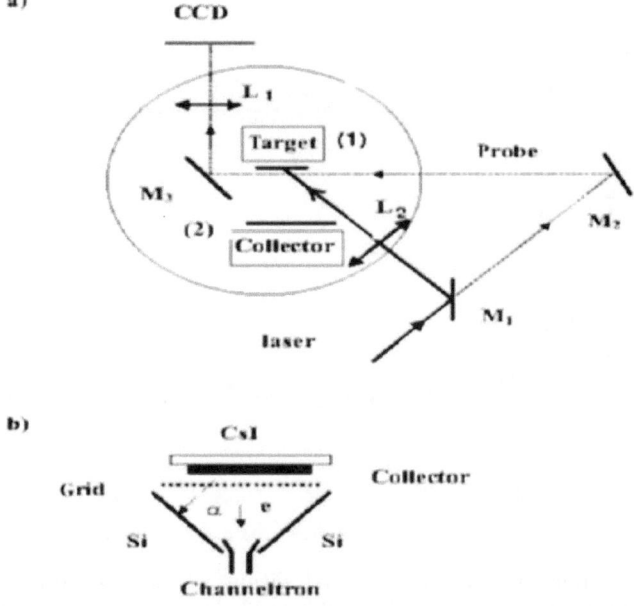

Fig. 13.2. (a) Schematic drawing of the setup in the irradiation position. L1 and L2 are focussing lenses; M1, M2, and M3 are mirrors. The difference in the path length between the main and the probe beams is adjustable by moving M2. The target and the collector are respectively mounted on the supports (1) and (2) which can be moved separately. At the end of the irradiation, the support (2) is automatically transferred in the detection chamber. (b) Schematic drawing of the detection setup installed under vacuum in a chamber not represented on the drawing. The collector and the Cesium Iodide detector are mounted on the same support. The collector is biased at 2 V. The thin dashed line represents a grid biased at 175 V. Two out of four Si detectors used in anticoincidence mode (to eliminate the delta electrons following the alpha decay of ^{234}U present in a very low concentration in the target) are represented. The Si detectors are mounted symmetrically with respect of the electron multiplier labelled channeltron. The other extremity of the channeltron, not represented, is biased at 2800 V. The efficiency of the detection setup is equal to $5.6 10^{-2}$

The difficulties encountered in this experiment due to exoelectron emission, self-absorption of the low-energy electrons in the catcher foil, and alpha particle emission by the radioactive U isotopes have been quantitatively evaluated with the consequence of a background signal considerably reduced with comparison with previous experiments on the same subject [8].

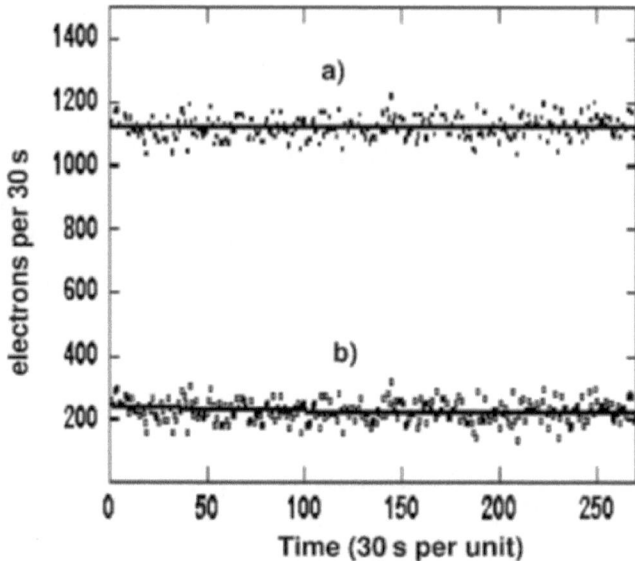

Fig. 13.3. Number of electrons detected within a time interval of 30 s versus the time elapsed after the end of the laser shots. The origin of the time on the figure is taken when the collector arrives in front of the electron detector. (**a**) The distribution corresponds to the sum of 10 independent measurements. (**b**) A distribution corresponding to (1) with $m_0 = 10$ has been added to the experimental distribution shown in (**a**) after subtraction of a constant value of 900 counts per channel for viewing purposes. The curve is the result of the fit

13.3 Excitation of an Isomeric State in ^{181}Ta

^{181}Ta has an excited isomeric state ($I = 9/2^-$, $T_{1/2} = 6.8\,\mu$s) located at 6.2 keV excitation energy. It decays via an E1 transition to the $I = 7/2^+$ ground state. The direct population of this state by the absorption of one photon is almost impossible to realize in view of the very small gamma width of the isomeric state ($\Gamma_\gamma = 6.7 \times 10^{-11}$ eV). However it has been reported recently by Andreev et al. [9] that in plasma conditions, created at laser intensities in the range of 1×10^{16} to 4×10^{16} W/cm^2, such an excitation has been observed at a rate of $(2 \pm 0.5) \times 10^4$ per laser shot! This would imply an increase of the nuclear level width of several orders of magnitude in the plasma environment ($\Gamma_\gamma \sim 0.3$ eV). Such a result, if confirmed, is of prime importance for the issue of energy storage in the nucleus.

The experiments are carried out at the Centre des Lasers Intenses et Applications at the University of Bordeaux-1. We have performed a precise characterization of the photon spectrum produced in the plasma resulting from the laser–solid Ta target interaction, in order to evaluate the nuclear excitation rate in our experimental conditions. We report here on the results of these measurements (collaboration CENBG, CELIA, CEA-DAM, IOQ Jena, the University of Strathclyde).

The ^{181}Ta plasma is produced with the Ti:sapphire CELIA laser system operating at a high-repetition rate of 1 kHz and delivering pulses with a wavelength of 800 nm. Pulse duration is compressed to 45 fs. The output energy can be changed from 0.4 to 2.2 mJ. Focussing the beam with a $f = 20$ cm lens, 78% of the laser energy is concentrated in a spot of 10-μm diameter, which allows to reach the intensity range scanned by Andreev et al. A schematic drawing of the experimental setup is shown in Fig. 13.4.

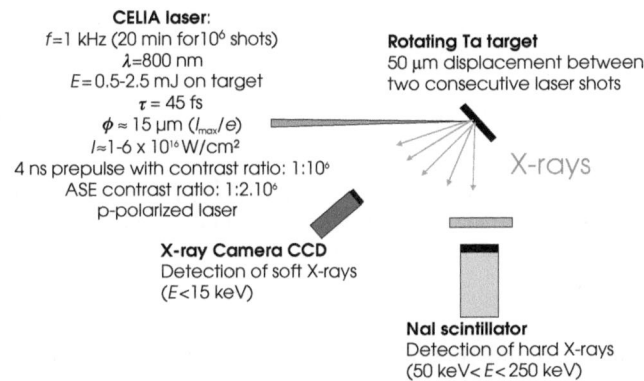

Fig. 13.4. Schematic representation of the experimental setup

A 5-mm-thick sodium Iodide, NaI(Tl), scintillator with a 300-μm Be window coupled to a photomultiplier tube is used to sample the hard X-ray spectrum. It is sensitive to X-rays with energy from 50 to 250 keV. It is surrounded with 5-mm-thick copper and lead layers. A 23-mm-diameter hole in the lead and copper tubes, centered on the NaI crystal, forms the entrance aperture for the X-rays emitted from the Tantalum plasma. A CCD X-ray camera was used to measure the lowest energy part of the photon spectrum (few keV – 20 keV).

In order to ensure single photon measurements, collimators and absorbers have been placed in front of the photon detectors. In the optimal experimental configuration, a signal was detected in the NaI detector every 10 shots at the most.

Examples of the photon spectra measured are shown in Fig. 13.5 for different laser intensities on target. These spectra are efficiency corrected. The response function of the detection setup has been calculated using the GEANT3 simulation code. From the number of photons produced in the 6 keV range we estimate, considering the natural width of the level, a maximum production rate of 0.01 isomeric state per laser shot.

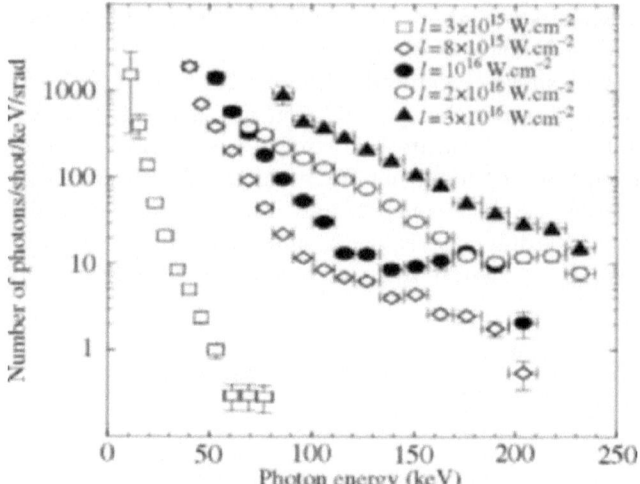

Fig. 13.5. Photon energy distributions as function of the laser intensity I. These spectra are unfolded with the NaI response function calculated with the GEANT3 simulation code which takes into account the presence of absorbers and collimators in front of the detector

This result confirms that the excitation rate per laser shot reported by Andreev et al. cannot be explained in terms of standard direct excitation of the isomeric state. We plan to measure this rate in a future experiment.

13.4 Effect of High Fields on Nuclear Level Properties

Several of the common modes of nuclear deexcitation proceed via an interaction between the nucleus and the electronic shells. Examples of these are the internal conversion process (IC) where the nucleus deexcites with emission of an electron and the $\beta\pm$-decay processes where electrons or positrons are emitted from the nucleus into the continuum. It is well known that the nuclear transition probabilities for these processes are strongly influenced by the Coulomb interaction and by the effects of screening due to the atomic electrons. These effects are particularly important in heavy nuclei.

The current generation of high-intensity lasers ($I > 10^{20}$ W/cm^2) are characterized by very strong electric fields, of the order of magnitude of those existing between atomic electrons and the nucleus (10^{11} V/m), in well-defined regions of space. These intense electric fields are capable of strongly modifying the electronic environment of the nucleus and therefore may considerably alter its decay properties. We would like to study the behavior of the internal conversion process in the presence of the laser field of the high-intensity lasers with wavelengths, of the order of the micrometer, available today. To our knowledge, this subject has not been investigated to date.

An obvious effect results from the ionization of the atom by the electromagnetic field. The change of the mean charge state of the atom will produce some modification of the electron-binding energies which will induce effects similar to those investigated in previous works on IC [1, 2], or $\beta^{+/-}$-decay [10]. In a serial of papers Kalman and collaborators [11] have discussed the influence of laser sources on the decay of the nucleus by IC. They have mainly investigated the possibilities offered by ultrashort X-ray wavelength lasers, delivering photons with energies of the order of magnitude of the nuclear transition energy, to trigger energetically forbidden IC decays. They call this process laser assisted IC decay. Experimentally, we want to produce nuclear excited states and study the modification of their deexcitation via IC in presence of a high-intensity laser field. In order to ensure that the laser field will apply on all the excited nuclei of interest, those should be produced in a subcritical plasma target. We propose to use three synchronized laser beams, the first one to produce a particle beam (of protons or heavy ions) with high enough energy to induce the nuclear reactions leading to the nuclear excited states of interest, the second one to produce the plasma target, and the third one to produce the high-intensity electromagnetic field. The choice of the nuclear reaction is limited by the impossibility today to detect nuclear prompt emissions in the presence of a laser beam due to the high level of background created in the target chamber. We will therefore produce isomeric nuclear states or β^+ emitters with lifetimes of the order of a few minutes in the plasma target. This plasma will be collected on a foil that will be placed, outside of the target chamber, in front of a counting station (composed of electron or photon detectors).

It is obvious that such an experiment cannot be performed with standard nuclear physics accelerators. The first stage of this program is planned in 2005 at the LULI facility at Polytechnique, Palaiseau (collaboration CENBG, LULI, CEA-DAM, IOQ Jena, the University of Strathclyde).

13.5 Conclusions

High-intensity lasers will offer in the future the opportunity to investigate nuclear properties in conditions that cannot be created with standard nuclear physics accelerators. For example, the possibility to focus several synchronized laser beams on the same target area will allow one day to perform nuclear reactions on excited states targets. In the meanwhile, an important effort is required to develop the detectors adapted to the laser conditions (huge particle fluxes of very short duration).

References

1. M. Morita: Prog. Theor. Phys. **49**, 1574 (1973)
2. F. Attallah, M. Aiche, J.F. Chemin, J.N. Scheurer, W.E. Meyerhof, J.P. Grandin, P. Aguer, G. Bogaert, J. Kiener, A. Lefebvre, J.P. Thibaud, and C. Grunberg: Phys. Rev. Lett. **75**, 1715 (1995), and T. Carreyre, M.R. Harston, M. Aiche, F. Bourgine, J.F. Chemin, G. Claverie, J.P. Goudour, J.N. Scheurer, F. Attallah, G. Bogaert, J. Kiener, A. Lefebvre, J. Durell, J.P. Grandin, W.E. Meyerhof, W. Phillips: Phys. Rev. C **62**, 24311 (2000)
3. M.R. Harston, J.F. Chemin: Phys. Rev. C **59**, 2462 (1999)
4. Y. Izawa, C. Yamanaka: Phys. Lett. **88B**, 59 (1979)
5. V.I. Goldansky, V.A. Namiot: Sov. J. Nucl. Phys. **33**, 169 (1981)
6. R.V. Arutyunyan et al.: Sov. J. Nucl. Phys. **53**, 23, (1991)
7. J.A. Bounds, P. Dyer: Phys. Rev. C **46**, 852 (1992)
8. G. Claverie, M. Aleonard, J.F. Chemin, F. Gobet, F. Hannachi, M. Harston, G. Malka, J. Scheurer, P. Morel, V. Méot: Phys. Rev. C **70**, 44303 (2004)
9. A.V. Andreev, R. Volkov, V. Gordienko, A. Dykhne, M. Kalashnikov, P. Mikheev, P. Nickles, A. Savel'ev, E. Tkalya, R. Chalykh, O. Chutko: J. Exp. Theor. Phys. **91**, 1163 (2000)
10. F. Bosh, T. Faestermann, J. Friese, F. Heine, P. Kienle, E. Wefers, K. Zeitelhack, K. Beckert, B. Franzke, O. Klepper, C. Kozhuharov, G. Menzel, R. Moshammer, F. Nolden, H. Reich, B. Schlitt, M. Steck, T. Stöhlker, T. Winkler, K. Takahashi: Phys. Rev. Lett. **77**, 5190 (1996)
11. T. Bükki, P. Kalman: Phys. Rev. C **57**, 3480 (1998), and references therein

14

Nuclear Physics with Laser Compton γ-Rays

T. Shizuma[1], M. Fujiwara[1,2], and T. Tajima[1]

[1] Kansai Advanced Photon Research Center, Kansai Research Establishment,
Japan Atomic Energy Research Institute, 8-2 Umemidai, Kizu 619-0215 Kyoto,
Japan
[2] Research Center for Nuclear Physics, Osaka Universy, Ibaraki 567-0047 Osaka,
Japan
shizuma@popsvr.tokai.jaeri.go.jp

Abstract. Photo-induced nuclear reactions are highly effective in studying nuclear physics and nuclear astrophysics. Photon beams created in laser Compton scattering have characteristics of monochromaticity, energy tunability, and high polarization. These photon beams are used in photo-nuclear experiments by nuclear resonance fluorescence (γ, γ') and photo-disintegration (γ, n) reactions. Recent experimental results and future plans are presented.

14.1 Introduction

A photon beam generated by inverse Compton scattering of laser photons (called laser Compton scattering [LCS]) with relativistic electrons has excellent characteristics of monochromaticity, energy tunability, and high polarization, and thus has provided unique research opportunities in nuclear physics and nuclear astrophysics.

In nuclear structure studies, nuclear resonance fluorescence (NRF) has been widely used [1]. In the NRF process, resonant states are excited by photo-absorption and are subsequently deexcited by photoemission. This method has an advantage that both the excitation and deexcitation take place through the electromagnetic interaction which is well analyzed on the theoretical basis only with ambiguities in nuclear structure calculation. The NRF measurements provide useful information on nuclear structures such as energies of the excited states, transition probabilities, nuclear spins, and parities. Circular-polarized photons which can be obtained by the laser Compton scattering are useful for measurements of parity nonconservation.

Photo-induced nuclear reactions also play a key role in understanding the nucleosynthesis of heavy proton-rich elements, the so-called p nuclei, which are produced by a series of photo-disintegration reactions of (γ, n), (γ, p), and (γ, α) at temperature of 2×10^9 to 3×10^9 K during supernova explosions [2, 3]. The product of the photon flux (given as the Planck distribution) and the

T. Shizuma et al.: *Nuclear Physics with Laser Compton γ-Rays*, Lect. Notes Phys. **694**, 217–229 (2006)

photo-disintegration cross section is proportional to the stellar reaction rate, leading to a narrow energy window above the threshold of particle emission with typical width of 1 MeV [4]. Detailed information on photo-disintegration cross sections at the low-energy tail of giant dipole resonance (GDR) is needed for the abundance calculation of the p nuclei.

Measurements on neutron capture cross sections of unstable nuclei are difficult in many cases. Inverse photo-disintegration reaction data may be used to estimate the neutron capture cross section on the basis of the reciprocity theorem within the statistical model calculation. This method is applicable to the s-process branching point nuclei whose β decay rates are the same order of magnitude as the neutron capture rates.

Photo-transmutation driven by intense laser plasma was recently reported [5, 6]. Photo-disintegration cross sections are large at the GDR region, and are nearly uniform as a function of the mass number. Photo-disintegration reactions on long-lived nuclei may transmute them to short-lived or stable nuclei.

In the next section, we describe photon beams used for experimental studies of nuclear physics and nuclear astrophysics. In Sect. 14.3, recent experimental results and proposed photo-nuclear experiments are presented. Nuclear transmutation by photo-nuclear reactions is discussed in Sect. 14.4. The summary is given in Sect. 14.5.

14.2 Laser Compton Scattering γ-Rays

In the early days of photo-nuclear experiments as described in [7], photons from radioactive isotopes produced by neutron capture reactions were used [8]. However, the limited energy of discrete γ-rays was the obstacle in measuring the photo-nuclear reaction. More recently, bremsstrahlung radiation by electron accelerators was utilized [9]. The bremsstrahlung radiation is produced by decelerating electrons in a massive radiator target, and has a continuous energy spectrum. Photons from positron annihilation in flight were also used for nuclear physics studies [10, 11]. Positrons impinging upon a thin, low-Z target create quasi-monochromatic, energy tunable annihilation photons. This photon beam, however, suffers from contamination of photons by positron bremsstrahlung radiation.

A different method to produce a high-quality γ-ray beam is Compton scattering of laser photons, the so-called LCS, with high-energy electrons. The LCS enhances the energy of the incident laser photons, in contrast to the conventional Compton scattering of photons with electrons at rest where the incident photon energy is consumed by recoiling the target electrons. In the case of head-on collisions of laser photons against relativistic electrons, the energy of scattered photons, E_γ, can be expressed as

$$E_\gamma \approx \frac{4\gamma^2 E_l}{1 + (\gamma\theta)^2 + 4\gamma E_l/mc^2} \, , \tag{14.1}$$

where E_l is the energy of the incident laser photon, γ is the relativistic Lorentz factor for the incident electron, θ is the scattered angle of the laser photon, and mc^2 is the electron rest mass. Most of the photons are scattered in the direction of the incident electron due to the momentum conservation. The technique to produce LCS photons using an electron storage ring was first suggested by Milburn [12].

14.2.1 LCS Photon Facility at AIST

A quasi-monochromatic γ-ray beam from laser Compton scattering has been developed at the National Institute of Advanced Industrial Science and Technology (AIST) [13]. A schematic overview of the AIST-LCS facility is shown in Fig. 14.1. The electron storage ring TERAS [14] provides electrons with $E_e = 200\sim750$ MeV. Together with an Nd:YLF laser at either primary ($\lambda = 1054$ nm) or second ($\lambda = 527$ nm) harmonics, photons with energies between 1 and 40 MeV can be obtained. A lead collimator located at downstream of the interaction area defines the scattering angle of photons ($\theta \approx 1/\gamma$) so that a quasi-monochromatic energy spectrum can be formed with a resolution of several % in FWHM. The electron beam divergence also affects the energy spectrum of LCS photons [15].

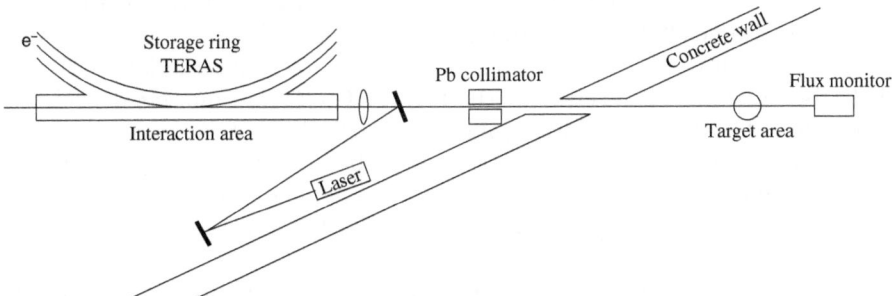

Fig. 14.1. A schematic overview of the AIST-LCS photon facility

In addition to the monochromaticity and energy tunability, high polarization is another advantage of the LCS γ-ray beam. So far, the polarization of more than 99% has been achieved at AIST [13]. The highly polarized photon beam is useful in nuclear structure studies for parity assignments of nuclear excited states.

14.2.2 New LCS γ-Ray Source

For the measurements of (γ, n) cross sections on rare isotopes or long-lived radioactive nuclei, a high-intensity γ-ray beam is needed for accumulating the good statistics data. In order to increase the flux of LCS photons, one of the promising methods is to use the free electron lasers (FEL) which employ relativistic electrons as the lasing medium to generate coherent radiation. Thanks to the recent developments of FEL, stable kW-level lasing has been realized [16, 17], and the power of FEL is expected to be further increased in near future. The characteristics of FEL such as high intensity, tunable wavelength, and sharp line width provide a benefit for photo-induced nuclear reaction experiments. LCS γ-ray beams based on intracavity Compton scattering of FEL photons have also been developed [18, 19].

14.3 Nuclear Physics and Nuclear Astrophysics

14.3.1 Nuclear Resonance Fluorescence Measurements: Parity Nonconservation

Parity nonconservation (PNC) is well known in nuclear physics after the discovery of the mirror symmetry violation in β decays [20, 21]. This mirror symmetry violation is now understood as the fundamental role of the weak bosons, $W^\pm$, which are mediators in β decays. In the past, the fundamental role in PNC was well studied via the β decay processes. Although the observations of the PNC effect in the nucleon–nucleon interaction are not quite new, the PNC studies in nuclear medium are not well understood.

A trial to observe the PNC effect in γ-decay processes started with the first report by Tanner [22], and followed by the work of Feynman and Gell-Mann [23] for the universal current–current theory of weak interaction. Wilkinson [24] also triggered the studies of the tiny PNC effect in nuclear deexcitation processes. The process contributing to the PNC effect is due to the weak meson–nucleon coupling between the weak bosons and meson exchanges in the N–N interaction (direct Z° weak bosons couple to the π, ρ, and ω mesons in the N–N vertex). The details of the PNC studies are reviewed in [25, 26, 27, 28].

It is concluded in [28] that the experimental PNC studies are still not satisfactory, and more studies from the experimental and theoretical sites are needed. Among many proposed experiments, one of clear experimental studies is to measure the parity mixing between the parity doublet levels. If there is parity mixing interaction between two very closely located states, the wave functions, $\tilde{\phi}_1$ and $\tilde{\phi}_2$, mixed by the PNC interaction V_{PNC} are obtained as
$$|\tilde{\phi}_1\rangle = |\phi_1\rangle + \frac{\langle \phi_2 | V_{\mathrm{PNC}} | \phi_1 \rangle}{E_2 - E_1} |\phi_2\rangle \text{ and } |\tilde{\phi}_2\rangle = |\phi_2\rangle + \frac{\langle \phi_1 | V_{\mathrm{PNC}} | \phi_2 \rangle}{E_1 - E_2} |\phi_1\rangle.$$

In the case of ^{21}Ne, the energy difference ΔE is only 5.7 keV and a large mixing is expected. Actually, $\langle V_{\mathrm{PNC}} \rangle$ was estimated as $V_{\mathrm{PNC}} \leq -0.029$ [29]. Similar examples exist in the $E1$ and $M1$ mixing transitions for ^{19}F($1/2^-$, 110 keV $\rightarrow 1/2^+$, g.s.), ^{18}F(0^-, 108 MeV $\rightarrow 1^+$, g.s.), and ^{175}Lu($9/2^-$, 396 keV $\rightarrow 7/2^+$, g.s.) [26, 27, 30]. In all the cases, the data accuracy is not so high, and the transition matrix is interpreted by a model calculation. In light nuclei, the

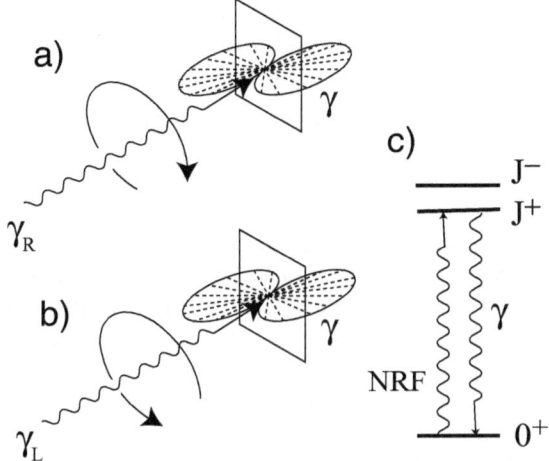

Fig. 14.2. Absorption of left- and right-handed circular polarized photons. Parity doublet levels are indicated

parity doublet levels could be well understood in the α-cluster model. Some problems in applying the shell model calculations are addressed in [27]. It is important to obtain results from a different type of experiment. We would like to point out one possibility to use a circularly polarized γ-ray beam.

Figure 14.2 shows a new scheme of the PNC measurement. The left- and right-handed circularly polarized photons are prepared by means of the inverse Compton scattering. Assuming that the photon beam is highly polarized, the difference of the photon absorption in the NRF process can be measured by changing the direction of the helicity of the photon beam. The $M1/E1$ mixing is expected to be the order of 10^{-3} to 10^{-7}. The difference of the NRF γ-ray yields $R = \frac{Y_L - Y_R}{Y_L - Y_R}$ is a direct measure of the parity nonconservation, which is proportional to the PNC matrix element M^{PNC} as $\sqrt{B(M1)/B(E1)}\cdot M^{\mathrm{PNC}}/\Delta E$ or $\sqrt{B(E1)/B(M1)}\cdot M^{\mathrm{PNC}}/\Delta E$.

One trial to measure the PNC effect has started with detecting the NRF γ-rays from the $1/2^-$, 110 keV first excited state in ^{19}F. This level can be excited using the intense photon beam from the Wiggler system at SPring-8 with an intensity of about 10^{13} photon/s and with a width of 0.1 keV at 110 keV. A feasibility test has been finished with a LiF crystal target with a thickness of 5 mm by using a Ge-detector. In this measurement, the NRF counts of the order of 10^{10} will be accumulated, which allows us to obtain more accurate PNC values than the previous experimental data of $-7.4 \pm 1.9 \times 10^{-5}$.

Another special case is the deuteron photo-disintegration and its inverse reaction $\boldsymbol{n} + p \rightarrow d + \gamma$. Since the wave function of deuteron is simple, the precise measurements of both the neutron capture reaction and deuteron photo-disintegration are theoretically analyzed to understand the coupling scheme of Z° boson to ω, ρ, and π mesons in the NN interaction [31].

14.3.2 Stellar Nucleosynthesis: Origin of the p Nuclei

Stellar nucleosynthesis is the term for the process of creating the chemical elements by nuclear reactions in stars. It is believed that the majority of nuclei heavier than iron were synthesized by two neutron capture processes called slow (s) and rapid (r) processes [32]. However, in the proton-rich side of the β stability line between ^{79}Se and ^{209}Bi, there exit 35 nuclei that cannot be synthesized by the s and r processes. These are called p nuclei. One of the production mechanisms of the p nuclei is a series of photo-disintegration (γ, n), (γ, p), and (γ, α) reactions on seed nuclei synthesized earlier by the s and r processes [33, 34]. The relevant process, therefore, is referred to as γ process, which takes place under the condition of the temperature $T = 2 \times 10^9$ to 3×10^9 K, density $\rho \approx 10^6$ g/cm^3, and time scale of the order of seconds. At present, the oxygen and neon-rich layers of massive stars during supernova explosions are considered to be the most promising site for the γ process [2, 3].

The modeling of γ-process nucleosynthesis requires an extended network calculation with more than 10,000 nuclear reactions involving both stable and unstable nuclei. The astrophysical nuclear reaction rates are inputs to this network calculation. While a large number of experimental data of (γ, n) cross sections at the GDR region are available, there exist few experimental data for astrophysically important energies that are located close above the neutron threshold with a typical width less than 1 MeV [4]. Therefore, most reaction rates have been derived from the statistical model calculations on the basis of the Hauser–Feshbach theory. In order to improve the determination of the reaction rates, threshold behavior of photo-disintegration cross sections has to be measured experimentally. Recently, photoneutron reaction rates in a stellar photon bath at a typical γ-process temperature were extracted by the superposition of bremsstrahlung spectra with different end-point energies [4, 35].

In recent photo-disintegration experiments at AIST, differential photo-disintegration cross sections at energies of astrophysical interest were measured by direct neutron counting using the monochromatic LCS γ-ray beam [36]. Detailed structures of low-energy GDR tails on ^{181}Ta [36] were revealed. The reaction product, that is, ^{180}Ta, is known as one of the p nuclei. The corresponding photo-disintegration cross sections near the threshold energy directly influence the γ-process production rates for these nuclei. The total stellar photo-disintegration rate includes the contribution of thermally excited states of which the reaction cross sections are not measured in many cases. Nevertheless, the photo-disintegration data on the ground state set strong constraints on the nuclear model parameters such as the $E1$ strength function, and thus help reduce uncertainties of the stellar rate predictions. The neuclosynthetic problem on the ^{180}Ta isomer will be discussed later.

14.3.3 s-Process Branching: Evaluation of Neutron Capture Cross Sections

The s-process nucleosynthesis takes place under the condition with a relatively low neutron density and low temperature where β decay rates are generally faster than neutron capture rates for nuclei along the path. This process tends to produce stable nuclei ascending the β stability line up to ^{209}Bi. However, in some cases where nuclei along the s-process path have long half-lifes of at least several weeks, the neutron capture can compete with the β decay. These unstable nuclei are called s-process branching point nuclei. The analysis of the s-process branching allows to determine neutron fluxes, temperatures, and densities relevant to the s process [37, 38]. An example of the reaction chain for the s and r processes is given in Fig. 14.3. Here, ^{185}W and ^{186}Re having the half-lifes of 75 and 3.8 days, respectively, are the s-process branching point nuclei.

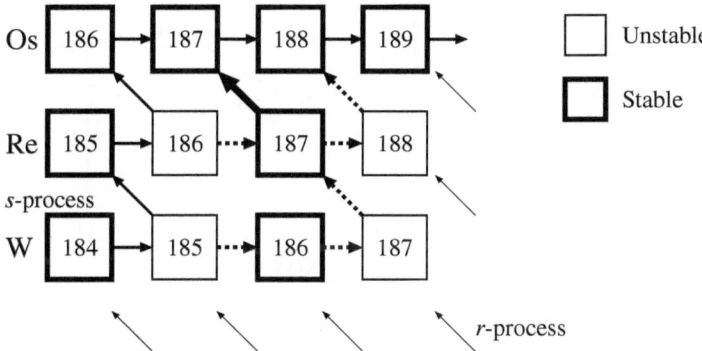

Fig. 14.3. The reaction chain for the element formation in the W–Re–Os region. The s- and r-process paths are shown with middle and thin *solid lines*, respectively. ^{185}W and ^{186}Re are the s-process branching points where the neutron captures shown with *broken lines* compete with the β decays. A *thick solid line* from ^{187}Re to ^{187}Os represents the cosmoradiogenic decay

Despite the progress of experimental techniques, it still remains extremely difficult to measure the neutron capture cross sections on short-lived nuclei such as the s-process branching nuclei ^{185}W and ^{186}Re. Instead, inverse photoneutron reactions may be used to estimate the neutron capture cross section on the basis of theoretical models [39, 40]. In this case, (γ, n) cross sections close to the threshold energies are important to constrain the model parameters.

In the following, neutron capture cross sections of ^{187}Os$(n, \gamma)^{188}$Os derived from the inverse photo-disintegration reaction ^{188}Os$(\gamma, n)^{187}$Os are compared with the measured values. Figure 14.4 shows measured photoneutron cross sections for ^{188}Os as a function of the average energy of the LCS γ-ray beam

[41]. The energy dependence of the photo-disintegration cross section was determined down to energies close to the neutron threshold (7.989 MeV). The ^{188}Os(γ, n) cross sections were calculated within the Hauser–Feshbach compound nucleus model using two different sets of input parameters, referred to as Calc. I and Calc. II [41]. The resulting ^{188}Os$(\gamma, n)^{187}$Os cross sections are compared with the experimental data in Fig. 14.4. It should be noted that the major difference between Calcs. I and II is attributed to the different treatment of the $E1$ strength function. Based on the parameter sets described above, the cross sections of the inverse neutron capture reaction on ^{187}Os were calculated. The results are compared with experimental data [43, 44] in Fig. 14.5. The neutron capture cross sections obtained from Calcs. I and II agree relatively well with the experimental data.

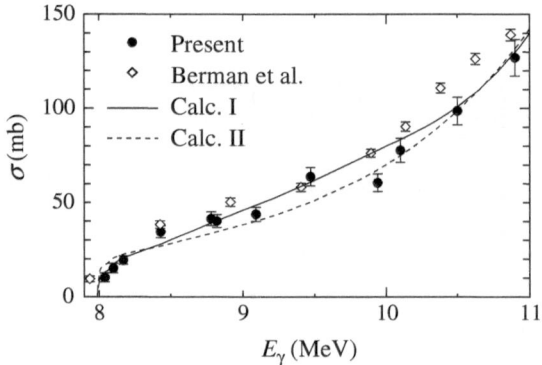

Fig. 14.4. Experimental ^{188}Os$(\gamma, n)^{187}$Os cross sections (*filled circles*) obtained in the present study. The data from [42] are also shown (*open diamonds*). The calculated cross sections are drawn by the *solid* (calc. I) and *dashed* (calc. II) lines

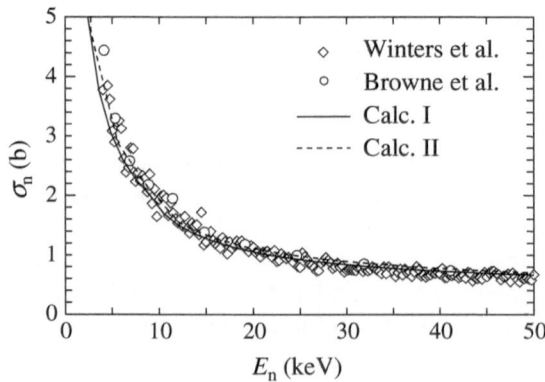

Fig. 14.5. Comparison between the calculated (*solid* and *dashed lines* for calcs. I and II, respectively) and measured (*open diamonds* from [43] and *open circles* from [44]) neutron capture cross sections on ^{187}Os

14.3.4 Deexcitation of the ^{180}Ta Isomer

The 9$^-$ isomer of ^{180}Ta (hereafter ^{180m}Ta where m denotes the meta stable state) is one of the most celebrating isomers. This isomer is famous for two aspects that it is the only naturally occurring isomer and the nature's rarest isotope. The ^{180m}Ta owes its existence to the highly K-forbidden transition between the 9$^-$ isomer and the 1$^+$ ground state (see Fig. 14.6). The half-life of ^{180m}Ta is more than 1.2×10^{15} years, while the ground state has a short half-life of $T_{1/2} = 8.1$ hours. The ^{180}Ta nucleus has received much attention in both nuclear structure physics [45] and nuclear astrophysics [36, 46]. We describe below a possible experiment of photo-induced deexcitation of the ^{180}Ta isomer relevant to stellar nucleosynthesis. The related photo-induced deexcitation of ^{180m}Ta through intermediate K-mixing states is depicted in Fig. 14.6.

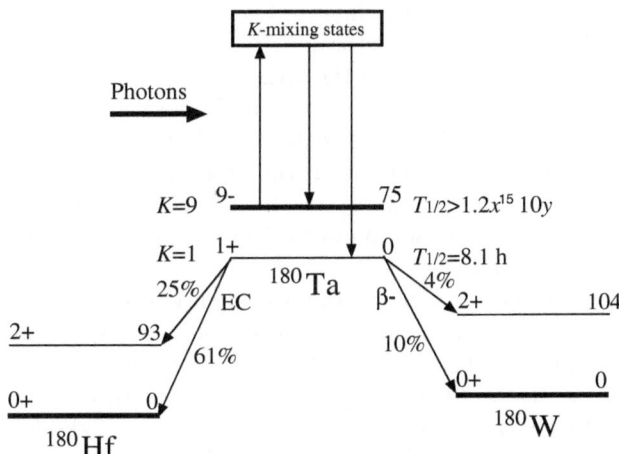

Fig. 14.6. Photo-induced deexcitation of the ^{180}Ta isomer via intermediate K-mixing states

The stellar production of ^{180m}Ta has been a challenging astrophysical problem, since the production mechanism of this isotope is still unknown. The production of ^{180}Ta is bypassed by the s process, and furthermore shielded from the β-decay chains following the r process. Possible ways to produce ^{180m}Ta have been proposed in terms of the γ-process path (see Sect. 14.3.2) and the s-process path [46]. In the latter case, ^{180m}Ta may be destroyed in (γ,γ') reactions under stellar plasmas at typical s-process temperatures. In the past, cross sections for the destruction of ^{180m}Ta by photons with energies higher than 1 MeV have been extensively measured [47, 48, 49]. However, the effects of intermediate K-mixing states lower than $E_x = 1$ MeV remain to be clarified which influences the effective half-life of ^{180}Ta [49]. The cross-sectional measurement for the depopulation of ^{180m}Ta with photon beam energies lower than 1 MeV could be important for its formation in the s-process nucleosynthesis.

14.4 Nuclear Transmutation

Photo-nuclear reactions of $(\gamma, 2n)$ and (γ, n) at GDR region have several unique features in view of their application to nuclear transmutation. The cross section, resonance energy, and width of photo-nuclear reactions depend little on the mass number, and have typical values of $\sigma \approx 0.4$ b, $E = 15 \sim 20$ MeV, and $\Gamma \approx 5$ MeV for $A = 100$ nuclei. Medium energy photons in the energy range of 10–30 MeV are effective for exciting GDR oscillations.

An efficacious way to generate a high-quality, large fluence photon beam with $E = 10 - 30$ MeV is via the inverse Compton scattering of copious laser photons off GeV electrons in a storage ring such as SPring-8. Usage of the electron storage ring has merits in the large fluence due to high current of circulating electrons. The photon scattered off the several GeV electrons spread no more than sub mrad. Electrons scattered off the laser photon lose energy by 10–30 MeV, but still can remain in the storage ring. They are soon reaccelerated up to the original energy in the RF cavity. Another merit is the low emittance of the resultant photons due to the low emittance of stored electrons and laser beam.

The energy of the photons in the 10–30 MeV range is spent in part to transmute nuclei via $(\gamma, 2n$ or $n)$ reactions, and in part to create electron–positron pairs. The neutron energy is typically a couple of MeV, and electrons and positrons are in the 10 MeV region. If the photon beam is well collimated, target nuclei to be transmuted can be confined in a cylinder with mm in diameter and sub m in length. The pair electrons produced by the 20 MeV photon are emitted forward with the average momentum of $P_e \sim 10$ MeV/c. Since they have transverse momentum of around 5–10 MeV/c, they are emitted outside the cylindrical target depositing little energy in the target. In short the photon energy is almost (more than 90%) converted into the electron kinetic energy, neutron biding energy, kinetic energy, and γ-ray energy.

When 5 kg of $A = 130$ nuclei corresponding to 3×10^{25} target nuclei are transmuted in a year, the photon flux of 2.5×10^{26}/year or 8.0×10^{18}/second are required using the target with 100 g/cm^2, and the average cross section of 0.2 b for the photons in the energy interval of 10–30 MeV. The numbers of electron–positron pairs and neutrons are around 2×10^{26} and 5×10^{25}, respectively.

14.5 Conclusion

Photon beams generated from LCS were used for studying nuclear physics and nuclear astrophysics. The characteristics of the LCS γ-ray beam such as monochromaticity, energy tunability, and high polarization is efficient in making the parity assignments of excited levels through nuclear resonance fluorescence (NRF) in nuclear structure studies as well as in measuring the photo-disintegration cross sections for nuclear astrophysics. The detailed structures of photo-disintegration cross sections near the threshold energies were measured for ^{93}Nb, ^{139}La, and ^{181}Ta for the γ-process study and ^{186}W, ^{187}Re, and ^{188}Os for the s-process study in nuclear astrophysics. The future experimental plan of photo-nuclear reactions using high-intensity LCS photons possibly realized by the free electron laser technology was presented.

Acknowledgment

Discussions with H. Utsunomiya, T. Hayakawa, H. Ohgaki, H. Ejiri, S. Goriely, and P. Mohr are gratefully acknowledged.

References

1. U. Kneissl, H.H. Pitz, A. Zilges: Prog. Part. Nucl. Phys. **37**, 349 (1996)
2. S.E. Woosely, W.M. Howard: Astrophys. J. Suppl. **36**, 285 (1978)
3. M. Rayet, M. Arnould, M. Hashimoto, N. Prantzos, K. Nomoto: Astron. Astrophys. **298**, 517 (1995)
4. P. Mohr, K. Vogt, M. Babilon, J. Enders, T. Hartmann, C. Hutter, T. Rauscher, S. Volz, A. Zilges: Phys. Lett. B **488**, 127 (2000)
5. J. Magill, H. Schwoerer, F. Ewald, J. Galy, R. Schenkel, R. Sauerbrey: Appl. Phys. B **77**, 387 (2003)
6. K.W.D. Ledingham, J. Magill, P. McKenna, J. Yang, J. Galy, R. Schenkel, J. Rebizant, T. McCanny, S. Shimizu, L. Robson, R.P. Singhal, M.S. Wei, S.P.D. Mangles, P. Nilson, K. Krushelnick, R.J. Clarke, P.A. Norreys: J. Phys. D: Appl. Phys. **36**, L79 (2003)
7. Handbook on photonuclear data for applications; Cross sections and spectra, IAEA-TECDOC-1178 (2000)
8. A. Wattenberg: Phys. Rev. **71**, 497 (1947)
9. M.K. Jakobson: Phys. Rev. **123**, 229 (1961)
10. C. Tzara: Compt. Rend. Acad. Sci. (Paris) **56**, 245 (1957)
11. B.L. Berman, S.C. Fultz: Rev. Mod. Phys. **47**, 713 (1975)
12. R.H. Milburn: Phys. Rev. Lett. **10**, 75 (1963)
13. H. Ohgaki, T. Noguchi, S. Sugiyama, T. Yamazaki, T. Mikado, M. Chiwaki, K. Yamada, R. Suzuki, N. Sei: Nucl. Inst. Meth. **A353**, 384 (1994)
14. T. Yamazaki, T. Noguchi, S. Sugiyama, T. Mikado, M. Chiwaki, T. Tomimasu: IEEE Trans. Nucl. Sci. **NS-32**, 3406 (1985)

15. H. Ohgaki, T. Noguchi, S. Sugiyama, T. Mikado, M. Chiwaki, K. Yamada, R. Suzuki, N. Sei, T. Ohdaira, T. Yamazaki: Nucl. Inst. Meth. **A375**, 602 (1996)
16. E.J. Minehara, M. Sawamura, R. Nagai, N. Kikuzawa, M. Sugimoto, R. Hajima, T. Shizuma, T. Yamauchi, N. Nishimori: Nucl. Instr. Meth. **A445**, 183 (2000)
17. G.R. Neil, C.L. Bohn, S.V. Benson, G. Biallas, D. Douglas, H.F. Dylla, R. Evans, J. Fugitt, A. Grippo, J. Gubeli, R. Hill, K. Jordan, R. Li, L. Merminga, P. Piot, J. Preble, M. Shinn, T. Siggins, R. Walker, B. Yunn: Phys. Rev. Lett. **84**, 662 (2000)
18. M. Hosaka, H. Hama, K. Kimura, J. Yamazaki, T. Kinoshita: Nucl. Inst. Meth. **A393**, 525 (1997)
19. V.N. Litvinenko, B. Burnham, S.H. Park, Y. Wu, R. Cataldo, M. Emamian, J. Faircloth, S. Goetz, N. Hower, J.M.J. Madey, J. Meyer, P. Morcombe, O. Oakeley, J. Patterson, G. Swift, P. Wang, I.V. Pinayev, M.G. Fedotov, N.G. Gavrilov, V.M. Popik, V.N. Repkov, L.G. Isaeva, G.N. Kulipanov, G.Ya. Kurkin, S.F. Mikhailov, A.N. Skrinsky, N.A. Vinokurov, P.D. Vobly, A. Lumpkin, B. Yang: Nucl. Inst. Meth. **A407**, 8 (1998)
20. T.D. Lee, C.N. Yang: Phys. Rev. **104**, 254 (1956)
21. C.S. Wu, E. Ambler, R.W. Hayward, D.D. Hoppes, R.P. Hudson: Phys. Rev. **105**, 1413 (1957)
22. N. Tanner: Phys. Rev. **107**, 1203 (1957)
23. R.P. Feynman, M. Gell-Mann: Phys. Rev. **109**, 193 (1958)
24. D.H. Wilkinson: Phys. Rev. **109**, 1603 (1958)
25. E.M. Henly: Annu. Rev. Nucl. Sci. **19**, 367 (1969)
26. E.G. Adelberger, W.C. Haxton: Ann. Rev. Nucl. Sci. **35**, 501 (1985)
27. B. Desplanques: Phys. Rep. **297**, 1 (1998)
28. W.C. Haxton, C.-P. Liu, M.J. Ramsey-Musolf: Phys. Rev. C **65**, 045502 (2002)
29. E.D. Earle, A.B. McDonald, E.G. Adelberger, K.A. Snover, H.E. Swanson, R. von Lintig, H.B. Mak, C.A. Barnes: Nucl. Phys. **A396**, 221c (1983)
30. B.R. Holstein: *Weak Interactions in Nuclei* (Princeton University Press, 1989)
31. M. Fujiwara: A.I. Titov, Phys. Rev. C **69** (2004) 065503
32. E. Margaret Burbidge, G.R. Burbidge, W.A. Fowler, F. Hoyle: Rev. Mod. Phys. **29**, 547 (1957)
33. D.L. Lambert: Astron. Astrophys. Rev. **3**, 201 (1992)
34. M. Arnould, S. Goriely: Phys. Rep. **384**, 1 (2003)
35. K. Vogt, P. Mohr, M. Babilon, J. Enders, T. Hartmann, C. Hutter, T. Rauscher, S. Volz, A. Zilges: Phys. Rev. C **63**, 055802 (2001)
36. H. Utsunomiya, H. Akimune, S. Goko, M. Ohta, H. Ueda, T. Yamagata, K. Yamasaki, H. Ohagaki, H. Toyokawa, Y.-W. Lui, T. Hayakawa, T. Shizuma, E. Khan, S. Goriely: Phys. Rev. C **67**, 015807 (2003)
37. F. Käppeler, H. Beer, K. Wisshak: Rep. Prog. Phys. **52**, 945 (1989)
38. F. Käppeler, S. Jaag, Z.Y. Bao, G. Reffo: Astron. Astrophys. **366**, 605 (1991)
39. K. Sonnabend, P. Mohr, K. Vogt, A. Zilges, A. Mengoni, T. Rauscher, H. Beer, F. Käppeler, R. Gallino: Astron. Astrophys. **583**, 506 (2003)
40. P. Mohr, T. Shizuma, H. Ueda, S. Goko, A. Makinaga, K.Y. Hara, T. Hayakawa, Y.-W. Lui, H. Ohgaki, H. Utsunomiya: Phys. Rev. C **69**, 032801(R) (2004)
41. T. Shizuma et al.: (in press)
42. B.L. Berman, D.D. Faul, R.A. Alvarez, P. Meyer, D.L. Olson: Phys. Rev. C **19**, 1205 (1979)
43. R.R. Winters, R.L. Macklin, J. Halperin: Phys. Rev. C **21**, 563 (1980)

44. J.C. Browne, B.L. Berman: Phys. Rev. C **23**, 1434 (1981)
45. G.D. Dracoulis, S.M. Mullins, A.P. Byrne, F.G. Kondev, T. Kibedi, S. Bayer, G.J. Lane, T.R. McGoran, P.M. Davidson: Phys. Rev. C **58**, 1444 (1998)
46. D. Belic, C. Arlandini, J. Besserer, J. de Boer, J.J. Carroll, J. Enders, T. Hartmann, F. Käppeler, H. Kaiser, U. Kneissl, M. Loewe, H.J. Maier, H. Maser, P. Mohr, P. von Neumann-Cosel, A. Nord, H.H. Pitz, A. Richter, M. Schumann, S. Volz, A. Zilges: Phys. Rev. Lett. **83**, 5242 (1999)
47. C.B. Collins, J.J. Carroll, T.W. Sinor, M.J. Byrd, D.G. Richmond, K.N. Taylor, M. Huber, N. Huxel, P. von Neumann-Cosel, A. Richter, C. Spieler, W. Ziegler: Phys. Rev. C **42**, R1813 (1990)
48. J.J. Carroll, M.J. Byrd, D.G. Richmond, T.W. Sinor, K.N. Taylor, W.L. Hodge, Y. Paiss, C.D. Eberhard, J.A. Anderson, C.B. Collins, E.C. Scarbrough, P.P. Antich, F.J. Agee, D. Davis, G.A. Huttlin, K.G. Kerris, M.S. Litz, D.A. Whittaker: Phys. Rev. C **43**, 1238 (1991)
49. D. Belic, C. Arlandini, J. Besserer, J. de Boer, J.J. Carroll, J. Enders, T. Hartmann, F. Käppeler, H. Kaiser, U. Kneissl, E. Kolbe, K. Langanke, M. Loewe, H.J. Maier, H. Maser, P. Mohr, P. von Neumann-Cosel, A. Nord, H.H. Pitz, A. Richter, M. Schumann, F.-K. Thielemann, S. Volz, A. Zilges: Phys. Rev. C **65**, 035801 (2002)

15

Status of Neutron Imaging

E.H. Lehmann

Spallation Neutron Source Division (ASQ), Paul Scherrer Institut, CH-5232
Villigen PSI, Switzerland
eberhard.lehmann@psi.ch

Abstract. This chapter describes the situation in the field of neutron imaging as a
tool for the investigation of macroscopic samples and objects. With the help of the
transmitted neutrons, providing a "shadow image" on a two-dimensional detection
system, a nondestructive analysis is possible. Although already in use since several
decades, the utilization of neutron-imaging techniques has become just more and
more important for practical applications due to many new aspects in the detector
development, improvements in the methodology on the one side, and the increased
requests to detect and to quantify, for example, hydrogenous materials in different matrixes on the other side. Compared to traditional film measurement common
some years ago, nowadays the digital methods have replaced it in most cases. There
are many new aspects from the physics side as phase contrast imaging, energy se-
lective imaging by using time-of-flight techniques, the use of pulsed sources, the
quantification of image data, and the access to fast neutrons in the MeV region.
The practical aspect of neutron imaging will be underlined by examples from the
author's work made together with his team in collaboration with several research
centers and with industrial partners. Based on these experiences, there is a reason for
optimism that neutron imaging will play an increasing role in the future in science
and technology.

15.1 Introduction

Neutrons as free particles interact with matter in different ways: by collision,
absorption, or even fission. Such reactions take place with the nuclei of the
involved materials, whereas the electron shell is out of interference.

Whereas the "production" of free neutrons is already a demanding process
(mainly by fission or spallation in special facilities), the utilization of neutron
beams for research and dedicated investigations has enabled to establish a
separate scientific field using sophisticated methods to study materials properties. This research area – neutron scattering – has increasing value for many
applications in solid state physics, soft matter analysis, and nuclear physics
in general. This has been the main reason to design and build intense neutron

E.H. Lehmann: *Status of Neutron Imaging*, Lect. Notes Phys. **694**, 231–249 (2006)
www.springerlink.com   © Springer-Verlag Berlin Heidelberg and European Communities 2006

sources with investments in the order of many hundred millions of euros and to exploit it with as many individual beam lines around the source as possible (e.g., FRM-2, SNS, J-PARC). From the physics point of view, slow neutrons are preferred because of the probability for both – the interaction with the sample material and the capture of the interacted neutrons in the detector, carrying the information about the sample after the collisions. Slow neutrons, so-called thermal or cold ones, have a wavelength (according to de Broglie's relation), which is in the order of the distances between the atoms in an atomic lattice. Therefore, the low-energy neutrons are the ideal probes to study the structure and the behavior of the sample material also in relative large size due to the fact that neutrons do not carry an electric charge enabling deeper penetration.

Whereas the scattered component in a neutron interaction is of major importance and interest for the field of neutron diffraction and neutron spectroscopy, neutron imaging is mainly dealing with the directly transmitted part of the beam. In the radiography mode, the transmitted neutrons are detected with a two-dimensional area neutron detector, which produces the "neutronic shadow image" of the object in the beam. All neutrons missing from the initial beam are considered as lost by attenuation in the object under investigation. When the neutron attenuation properties (i.e., the total interaction cross section) of the object are known, the material quantity can be obtained from the image data in principle. The problems occurring from such simplified considerations will be described in more detail below.

Neutron imaging techniques (radiography, tomography, real-time imaging, laminography,...) can derive similar results as common in X-ray imaging. The difference between both techniques is given by the interaction mechanism: neutrons interact with the atomic nucleus, X-ray with the electronic shell. Whereas the interaction probability for X-ray is correlated to the number of electrons in the shell and therefore with the mass number, the situation is completely different for neutrons. Light elements as hydrogen, lithium, and boron deliver high contrast for neutrons, but are more or less transparent for X-ray.

Contrarily, heavy materials like lead, bismuth, or even uranium are relatively transparent for neutrons but never for X-ray. On the basis of this behavior of the different radiation, it is obvious that both methods are complementary to each other, without specific preference for one of them. It depends very much on the topical problem which method is best suited for a nondestructive description of the assembly. A simple picture of the interaction probabilities is given with the periodic tables in Figs. 15.1 and 15.2, where the grey level indicates the attenuation ability of the natural isotope mixtures. The attenuation coefficients in numbers are given in these figures too.

Certainly, neutron imaging is more exotic as that with X-ray due to the availability of suited neutron sources. In all relevant cases, the size of the source makes the method of neutron imaging more stationary than a mobile one. In this chapter, the aspects for state-of-the-art strategies, facilities,

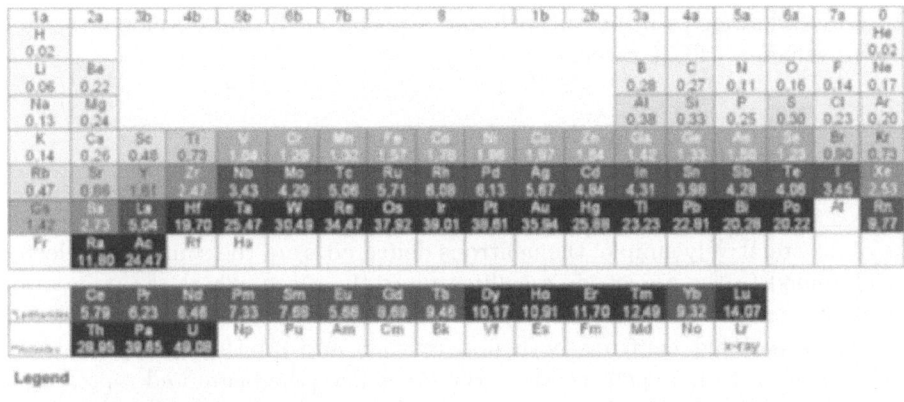

Legend

Attenuation coefficient [cm?1]= sp.gr.* µ/δ

sp.gr.: *Handbook of Chemistry and Physocs,* 56th edition 1975–1976.
µ/δ: J. H. Hubbell+ and S. M. Seltzer Ionizing Radiation Division, Physics Laboratory National Institute of Standards and
   Technology Gaithersburg, MD 20899.
   http://physics.nist.gov/PhysRefData/XrayMassCoef/tab3.html

Fig. 15.1. Periodic table of the elements with the attenuation coefficient for X-ray at 120 kV

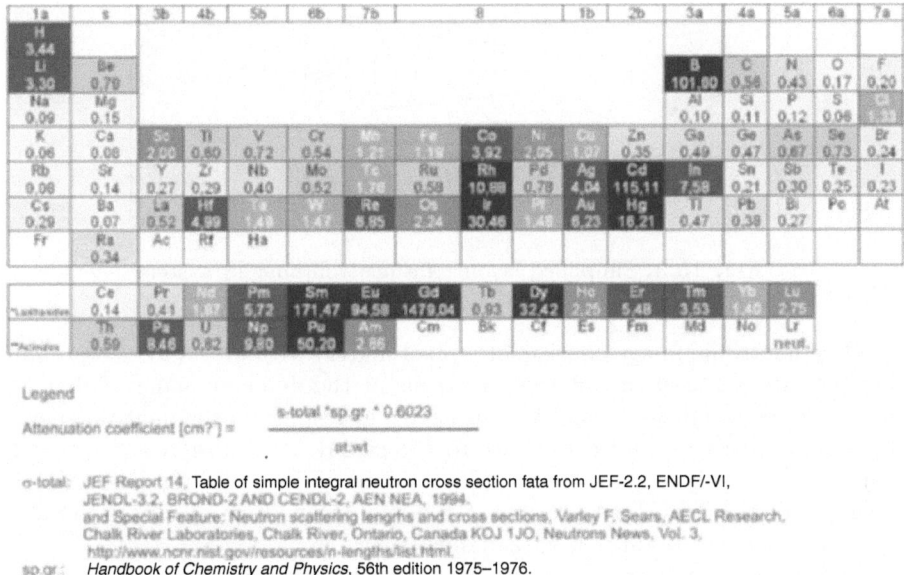

Legend

$$\text{Attenuation coefficient [cm?]} = \frac{\sigma\text{-total *sp.gr.* * 0.6023}}{\text{at.wt}}$$

σ-total: JEF Report 14. Table of simple integral neutron cross section fata from JEF-2.2, ENDF/-VI,
   JENOL-3.2, BROND-2 AND CENDL-2, AEN NEA, 1994.
   and Special Feature: Neutron scattering lengrhs and cross sections. Varley F. Sears. AECL Research.
   Chalk River Laboratories, Chalk River, Ontario, Canada K0J 1J0, Neutrons News. Vol. 3.
   http://www.ncnr.nist.gov/resources/n-lengths/list.html
sp.gr: *Handbook of Chemistry and Physics,* 56th edition 1975–1976.
at.wt.: *Handbook of Chemistry and Physics,* 56th edition 1975–1976.

Fig. 15.2. Periodic table of the elements with the attenuation coefficient for thermal neutrons

and investigations will be outlined and a vision for future developments and improvements will be given.

15.2 The Setup of Neutron Imaging Facilities

As sketched in Fig. 15.3, the principal arrangement for a neutron imaging facility looks relatively simple: the neutrons delivered from the source are selected and guided to the object via a collimator to the place, where the interaction with the sample material takes place. The detector behind the object registers all arriving neutrons both unperturbed and interfered by the object. The detector is arranged mostly perpendicular to the beam and represents a two-dimensional array of image dots (pixels).

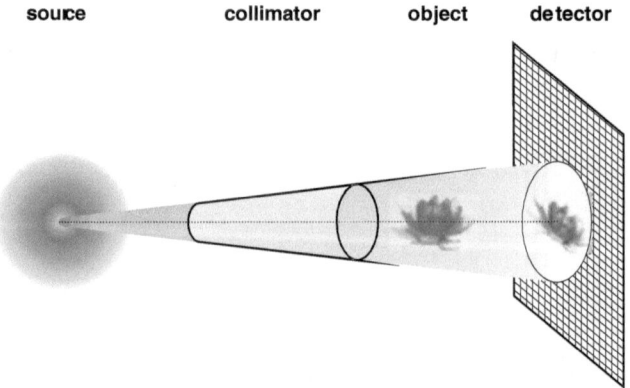

source   **collimator**   **object**   **detector**

Fig. 15.3. Simplified layout of a neutron imaging system

The neutronic image consists of a pixel matrix with grey values representing the intensities of the arriving neutrons at the detector plane. It depends on the detector efficiency and its sensitivity in respect to the neutron energy, how many neutrons will contribute to the signal. In reality, a neutron radiography facility is much more complex as demonstrated in Fig. 15.4 with the example of the NEUTRA station at the spallation neutron source SINQ at the Paul Scherrer Institut (Switzerland).

A major boundary condition for a neutron radiography system is the shielding around to satisfy the radiation protection regulations. The direct neutron beam, the accompanying gamma radiation, the secondary radiation delivered from the interaction with the sample, and the beam dump have to be considered. The access to such a shielded facility, in most cases made of concrete, is to control by protective measures. Special shutter devices guarantee that only that neutron field is applied, which is needed for the investigations. This is especially important for neutron beams to avoid extra activation

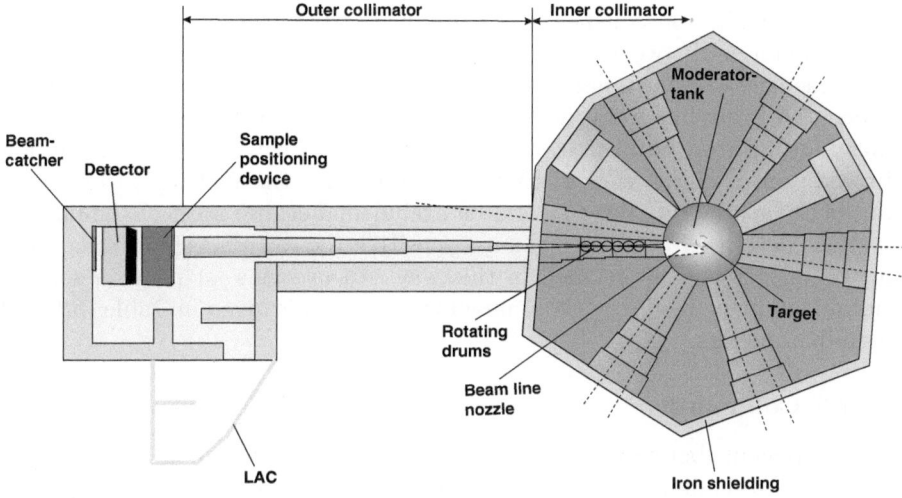

Neutra

Fig. 15.4. Top view onto the station for neutron imaging NEUTRA at the spallation source SINQ; the distance from the target in the center of SINQ to the outer wall of the shielding in beam direction is about 15 m

in construction materials. The performance of a radiography system is determined by all components: source, collimator, and detector. In the consequence, the setup conditions will define which kind of samples is able to be inspected preferably. The influence of the system components onto the performance is discussed in detail as follows.

15.2.1 Source

Powerful neutron sources are either fission reactor based or accelerator driven, so-called spallation neutron sources. The majority of neutron radiography stations are located at reactors; only the home facility of the author utilizes the thermal neutrons from the world strongest spallation source SINQ [1]. The image quality as described in this report will never be obtained by mobile sources as the radioisotope-driven ones due to the lack of intensity and collimation. As mentioned above, thermal or even cold neutrons are preferred for imaging purposes. Therefore, the moderation process for slowing down the initial fast neutrons from the nuclear reaction plays an important role for the beam quality obtained from the neutrons from the initial source region. Two moderator materials are in practical use: light water and heavy water. Despite to the cost aspect (D_2O is relatively expensive), heavy water is to prefer because of the loss-free dissipation of the thermalized neutron in an extended volume. More efficient extraction of the neutrons from the source to the beam lines is then the consequence. Light water absorbs much more neutrons than heavy water with two results: the rapid loss of intensity in larger distance from the core and the emission of gamma radiation from the neutron capture process. A beam line from the D_2O moderated source is therefore to prefer for imaging purposes because of the low contamination with gamma radiation. In addition, a direct view of that beam line onto the fuel region of a reactor or the spallation target should be excluded. Because most of the imaging systems described below are gamma sensitive too, an undesired background in the images can be avoided in this way. An overview of neutron sources, suitable in principle for neutron imaging purposes, is given in Table 15.1 with main parameters.

15.2.2 Collimator

All components between the primary source point and the sample position are considered as collimator. It can contain filters for the reduction of gamma and fast neutron background, limiters to reduce the beam size, and also shutters to enable the time-dependent beam supply. It is the aim of the collimator to deliver a quasi-parallel clean beam to the object under investigation with highest possible neutron intensity. Therefore, it is ever a trade-off between the beam collimation (expressed by the L/D-ratio) and the neutron flux level at the detector plane. The higher the beam intensity the lower will be the exposure time, that is, the higher the image frame rate. The beam collimation can influence the spatial resolution in the images when the object is in a certain distance d from the image plane. The geometric unsharpness U_g is directly linked to the collimation ratio as:

$$U_g = \frac{d}{L/D}.$$
(15.1)

Typical values for L/D are between 100 and 1000 and the spatial resolution is limited therefore as in the range of about 10 to 100 μm, depending on the sample geometry. However, there are some other effects limiting the spatial resolution as discussed below. A comparison to synchrotron radiation conditions is given by Fig. 15.5, describing the range of work in respect to sample size and spatial resolution. It becomes clear that neutron imaging is preferably to apply on the macroscopic scale from 0.1 mm to 10 cm due to higher transmission and the limitations in the spatial resolution by different reasons.

A magnification procedure as possible with microfocus X-ray tubes is impossible for neutrons due to missing sources with adequate intensity. Only an ideally parallel beam can be obtained as an optimum in respect to spatial resolution.

15.2.3 Detectors for Neutron Imaging

Neutrons cannot be detected directly without been converted to ionizing radiation, which makes then the real excitation process. Strong neutron absorbers,

Table 15.1. Overview about neutron sources for imaging purposes

Source Type	Nuclear Reactor	Neutron Generator	Spallation Source	Radio Isotope
Reaction	Fission	D-T fusion	Spallation by protons	γ-n-reaction
Used material	U-235	Deuterium, tritium	High mass nuclides	Sb, Be
Gain				
Primary neutron intensity (1/s)	1.00E+16	4.00E+11	1.00E+15	1.00E+08
Beam intensity (1/cm^{2}s)	10^6 to 10^9	10^5	10^6 to $n* \, 10^7$	10^3
Neutron energy	Fast, thermal and cold	Fast, thermal	Fast, thermal and cold	24 keV, thermal
Limitation of use	Burn up	Lifetime tube	Target lifetime	Half-life Sb-124
Typical operation cycle	1 month	1000 h	1 year	0.5 year
Costs of the facility	High	Medium	Very high	Low

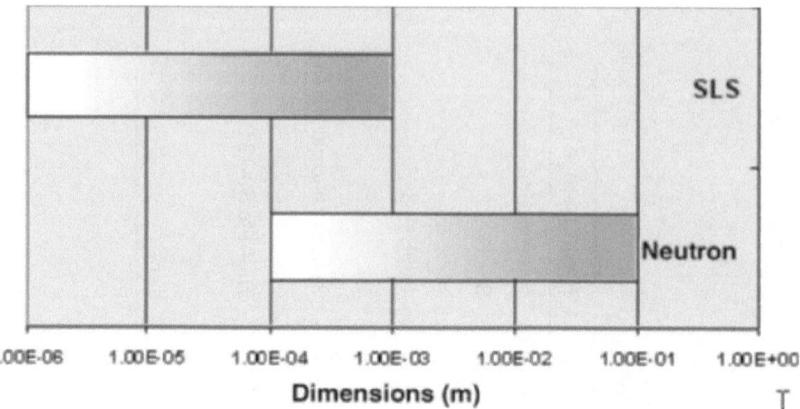

Fig. 15.5. Dimensions of samples to observe with either neutrons or synchrotron light (SLS), given by the attenuation behavior of the sample and the beam size, respectively

as Gd, Li-6, or B-10 are, favorably be used in several imaging detectors were the resulting information after the neutron capture is given via light emission, electric excitation, or a chemical reaction. For a long time until the eighties of last century, X-ray film in conjunction with a converter foil was the only significant system in routine use for neutron imaging purposes. The development in the recent years delivered a variety of new imaging systems, mainly based on electronic devices. As demonstrated in Fig. 15.6, the new systems cannot deliver the superior spatial resolution of film, but provide a lot of other advantages as much higher sensitivity, wide dynamic range, linearity over the full range, and an output in digital format, enabling the application of all features of image postprocessing.

The progress on the detector side as illustrated in Fig. 15.6 has consequences in the performance and the application range on neutron imaging systems in respect to new applications in research and industry. This process is not yet completed because new options become feasible to increase the spatial and time resolution.

15.3 Modern Neutron Imaging Detectors

It is worth to emphasize the new situation in respect to the neutron imaging detectors, which have brought a completely new situation in the field. As mentioned before, the advantages in the performance and in the utilization of neutrons generally enable new imaging methods fields as neutron tomography, phase contrast imaging, real-time studies, and laminography. Some of these techniques will be described below in more detail.

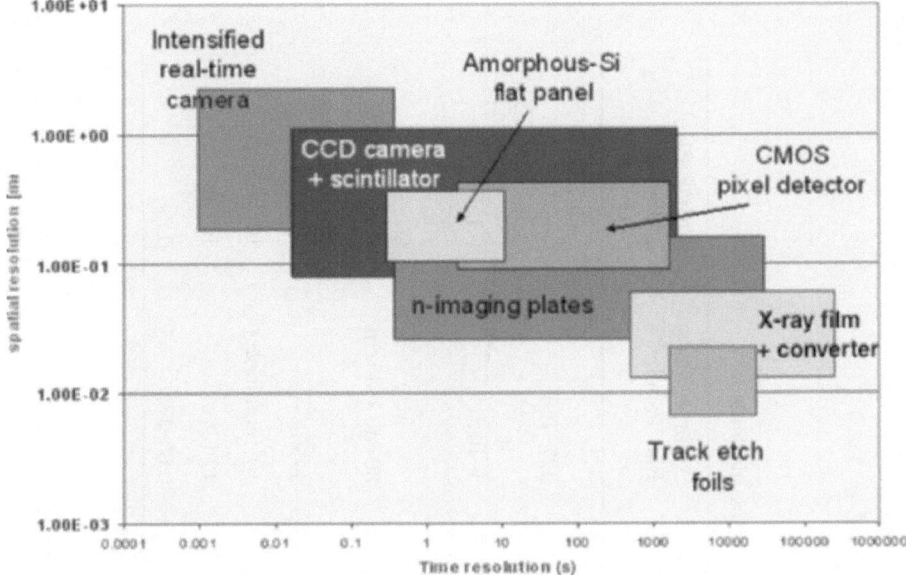

Fig. 15.6. Working area of neutron imaging detectors in respect to their time and spatial resolution, given for the conditions provided at the neutron radiography station NEUTRA at PSI. The dynamic range of the imaging system is defined by the length of the bar in respect to the time scale

An overview about digital neutron imaging devices is given in Table 15.2. A replacement of film methods is given by *Imaging Plates* using the radiation induced photo-stimulated luminescence to generate digital image data. The wide dynamic range, the high sensitivity, and the option to erase and reuse the plates enable very fast and flexible investigations for customers. In the case of neutron imaging, the converter material Gd is directly mixed into the sensitive layer as a modification of standard plates used in medical applications.

A powerful and flexible setup for digital neutron imaging is given by *CCD-camera detectors* observing the light excitation of a neutron-sensitive scintillator. These devices are in use in different setups at several beam lines worldwide. The main need is to find a very light sensitive camera operating on very low noise level over longer exposure time (seconds to minutes). Although expensive, these devices are nowadays the mostly applied systems for standard radiography and tomography applications as well.

If the light output from the scintillator is not enough to describe the investigated process in the relevant time, a light magnification can be obtained by *intensifiers*, mostly based on microchannel plate devices. With this technique, processes in the time scale of milli- or even microseconds can be observed. However, the image quality will be reduced in such cases because the noise level will be magnified too accordingly.

Table 15.2. Properties of digital neutron imaging methods

Detector System	X-ray Film and Transmission Light Scanner	Scintillator+ CCD-camera	Imaging Plates	Amorph Silicon Flat Panel	CMOS Pixel Detector
Max. spatial resolution [μm]	20–50	100–500	25–100	127–750	50–200
Typical exposure time for suitable image	5 min	10 s	20 s	1–10 s	0.1–50 s
Detector area typical	18 cm × 24 cm	25 cm × 25 cm	20 cm × 40 cm	30 cm × 40 cm	3.5 cm × 8 cm
Number of pixels per line	4000	1000	6000	1750	400
Dynamic range	10^2(nonlinear)	10^5(linear)	10^5(linear)	10^3(nonlinear)	10^5(linear)
Digital format	8 bit	16 bit	16 bit	12 bit	16 bit
Readout time	20 min	2–100 s	5 min	0.03–1 s	0.2 s

Recently, the *amorphous silicon flat panel* technology became available also for neutrons. These devices are placed in the direct beam because the radiation damage is considered here less important than for single crystals. Because of the high sensitivity and the fast readout, sequences up to 30 frames per second become possible [2, 3]. There is only small experience about the performance and long-term stability of these systems under permanent neutron bombardment.

The family of electronic detectors for neutron imaging purposes is completed by approaches to use the *CMOS technology* for a direct conversion of radiation into charge. In the case of neutron detection, prior to the radiation measurement in individual pixels of the CMOS chip, a conversion to ionizing radiation by neutron capture is needed. Special designed CMOS arrays with integrated amplifiers and counters per individual pixel enable a direct counting and thresholding. This development is not yet completed but promising.

15.4 Improved Neutron Imaging Methods

15.4.1 Radiography

Because of the high efficiency of state-of-the-art imaging detectors, a single frame image can be obtained within few seconds. The field of view depends on the beam diameter and the detector area, typically 20–40 cm in diameter. By a scanning routine and adding of individual frames, much larger objects can be observed as helicopter blades [4] or automotive car components [5]. With the help of image postprocessing tools, the valid dynamic range of the transmission experiment can be adapted to human visibility. Storage, archiving, and data transfer based on the digital information are no problem anymore.

15.4.2 Tomography

To enable the observation of macroscopic samples, in all three dimensions, tomography methods can be applied. In the case of neutron beams, usually a parallel one is presumed. This enables the application of relatively easy reconstruction algorithm based on filtered back-projection, performing the Inverse Radon Transformation:

$$\sum(x,y) = \int_0^\pi P(x \cdot \cos(\theta) + y \cdot \sin(\theta), \theta) . \, d\theta \qquad (15.2)$$

The reconstruction is performed in the x, y-plane, whereas the data in the third dimension are stacked layer by layer. The resulting array of attenuation coefficients $\sum$ for each volume element (voxel) has been determined by many individual frames of the object from different positions (rotation angle θ around the vertical rotation axis) – projections P. Depending on sample size

Fig. 15.7. Neutron tomography images of a sprinkler nozzle – outer surface, central slice, segmented O-ring. The object has a length of about 5 cm. The liquid in the glass capsule and the rubber ring gives especially high contrast for neurons compared to the metallic structure

and the requirements for spatial resolution, the number of projections should be between 200 and 1000. In this way, the generation of one volume data set needs between 0.2 and some hours per sample. An example of a resulting view of the reconstructed and visualized object obtained with neutron tomography is given in Fig. 15.7.

15.4.3 Quantification

A neutron transmission image represents the distribution of the neutron intensity $I(x, y)$ in two dimensions, while integrating over the third one. When the initial distribution $I_0(x, y)$ is known, the transmitted beam can be described according to

$$I(x, y) = I_0(x, y) \cdot \int e^{-\sum (x,y,z)} \, \mathrm{d}z \ . \tag{15.3}$$

This exponential law of attenuation is valid in first order for small sample thickness d or small macroscopic cross sections S, and for monoenergetic neutron beams. Then it becomes possible by inverting (15.3) to derive either the material concentration if the thickness is known or the material thickness if the material composition is known. Such kind of studies exploit the advantage of digital neutron imaging that $I(x, y)$ and $I_0(x, y)$ can directly be derived from the image data. In this way, each digital neutron image has to be considered as a data set in first order, describing the neutron transmission process. For thicker material layers and strong neutron scattering materials (as hydrogen and steel) the relation becomes misleading by underestimation of the attenuation. This is caused by the fact that scattered neutrons contribute to the resulting image when they are detected beside the direct line of response. To overcome and correct this problem, methodical studies have been performed on the basis of Monte Carlo simulation of the neutron transport process [6].

There will be tools available to solve the problem caused by multiple neutron scattering for both radiography and tomography in the near future [7].

15.4.4 Real-Time Imaging

Many requests in research and industrial applications need a time resolution in the observations. The time scale can be a sequence of images over hours to days, repetitive processes with many cycles per minute, or a very rapid but single event. For such kind of investigations, special options of the available detection systems have to be exploited. Generally, there is a limitation in time resolution by the neutron intensity. The strongest beam available for imaging purposes (Neutrograph at ILL [8]) has flux intensity of about $10^9 \, \mathrm{cm^{-2} s^{-1}}$, which enables exposure time of few milliseconds per frame. At facilities with less intensity, the option of triggering can be applied if repetitive processes will be investigated. By synchronizing the CCD with the running process and internal integration of as many frames as possible interesting results can be obtained, for example, the oil distribution in running engines (see Fig. 15.8 left). Another example (Fig. 15.8 right) on more relaxed time scale is the in situ investigation of plants growing under certain boundary conditions that can be achieved only with neutrons in such precision due to the high sensitivity to hydrogenous substances as plants.

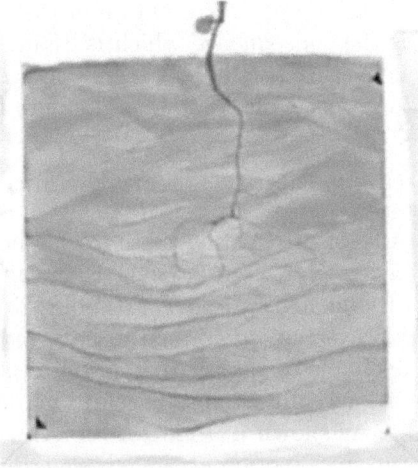

Fig. 15.8. Head of a motor cycle engine (*left*) where the redistribution of the lubricant can be visualized on piston motion frequencies of several thousand rotations per minute. On much slower time scale is the growing of roots (*right*), where the assembly is investigated under identical conditions over weeks with different moistening conditions

15.4.5 Phase Contrast Enhanced Imaging

Neutrons can also be considered as waves corresponding to the de Broglie relation. In this context a refraction index can be derived for a material in interaction:

$$n(x, y, z, \lambda) = 1 - \delta(x, y, z, \beta) - i\beta(x, y, z, \lambda) \, . \tag{15.4}$$

The parameter β represents the absorption properties, but δ the phase contribution in the interaction with matter. However, δ is very small – in the order of 10^{-6} – and not comparable with numbers for visible light.

To exploit the phase shift properties at material boundaries, a field of spatial coherent neutrons is required. This can be obtained by very small apertures far from the point of investigations. Wave fronts are considered as transversal coherent when the following relation is satisfied:

$$l_t < \frac{r \cdot \lambda}{s} \, , \tag{15.5}$$

whereas the wavelength λ, the distance between the source and the object r, and the source dimension s have to take into account. With $\lambda = 1.8 \,\text{Å}$ for thermal neutrons, a distance r of about 6 m and pinholes in the order of 0.5 mm, the coherence length will be in the order of some micrometers. In this way, the effects at boundaries become important because they are much better than the spatial resolution of the detectors systems. A contrast enhancement is the consequence which is especially important for weak neutron absorbers delivering small absorption contrast. An example of such kind of inspections is given in Fig. 15.9 for aluminum foam.

15.4.6 Energy Selective Neutron Imaging

Almost all existing facilities for neutron imaging purposes are using a poly-energetic ("white") beam that can be approximated by a Maxwell function around a mean energy (about 25 meV for thermal neutrons). It would be, however, an advantage to have monoenergetic neutrons with variable energy available to exploit the option to perform investigations near the Bragg edges of some important structural materials.

As shown in Fig. 15.10 for the example of iron, there is a strong slope in the cross section at the wavelength of 4 Å by a factor of more than 3, enabling big contrast in measurements above and below that Bragg edge.

In order to provide suited beams with narrow energy spectrum in this range, there are two options:

- A turbine-type energy selector, in use at different beam lines for neutron scattering experiments at cold guides
- To use a time-of-flight technique at a pulsed neutron sources

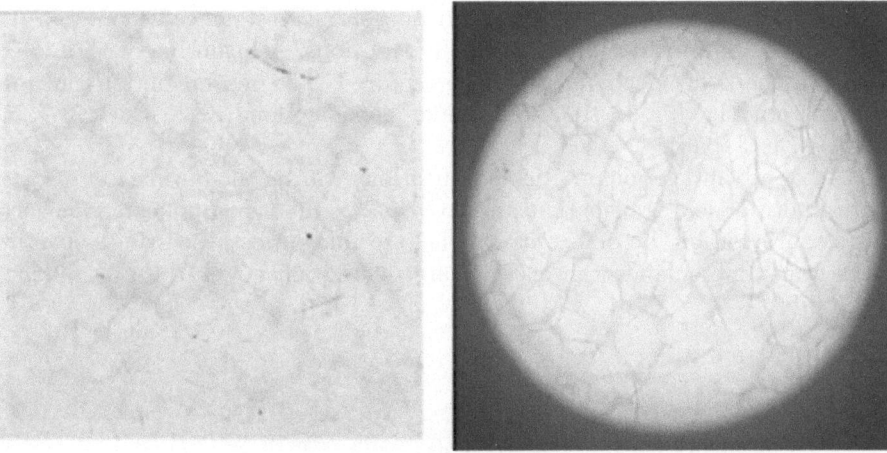

Fig. 15.9. The investigation of aluminum foam with transmission neutron imaging (*left*) and phase enhanced imaging (*right*), where an enhancement of the structure becomes visible

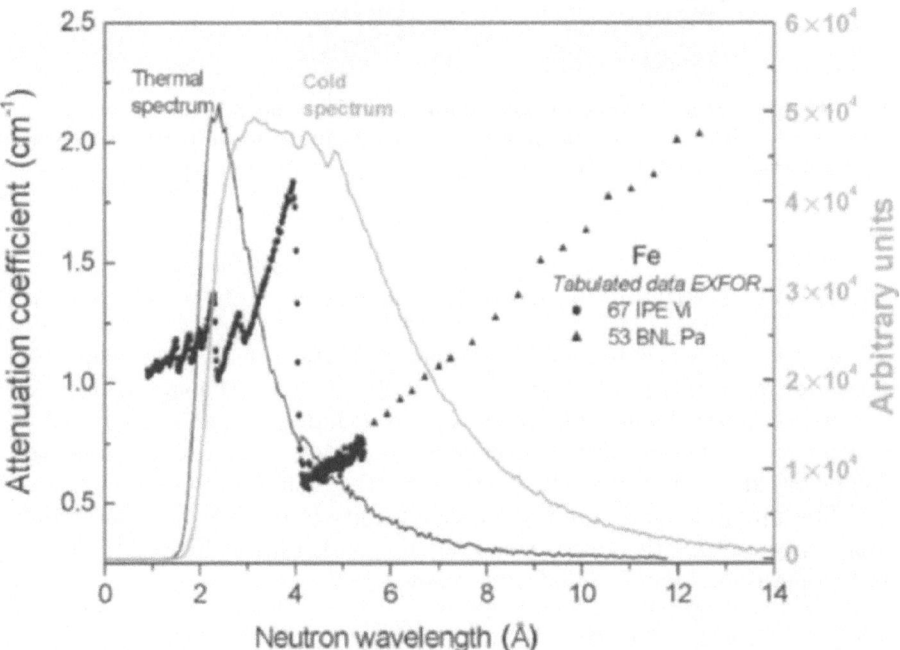

Fig. 15.10. The iron cross section in the wavelength range of 4 Å shows a strong slope (Bragg edge), which can be used for contrast enhancement when neutrons are applied near this energy for imaging

Both options are demanding, but promising. First, tests have been performed successfully [9], but not under optimal conditions. A beam line with a maximum in the cold spectrum would be required. The second option can take profit from the big installations under construction (SNS in U.S.A. and J-PARC in Japan).

A big advantage for practical applications will be the option to set structures more or less transparent and to visualize inner components efficiently. This can favorably be done with the help of image postprocessing, when images from different neutron energies are related as shown in the example in Fig. 15.11.

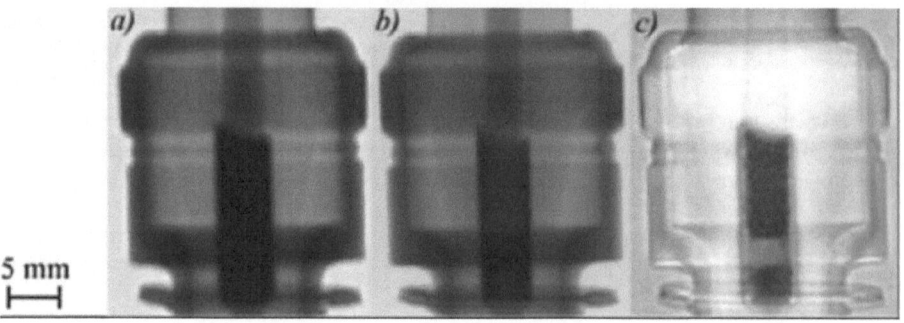

Fig. 15.11. Images of a spark plug obtained at 6.9 and 3.2 Å, respectively. The third image corresponds to the division of both, enabling much better visualization of the electrode isolation. Taken from [9]

15.4.7 Fast Neutrons for Imaging Purposes

When the sample size becomes in the order of 10 cm or more of compact material, for most of the objects it becomes difficult to have enough transmission with thermal or cold neutrons. There is only a chance for penetration with fast neutrons with energies in the order of about 1 MeV. Such sources of neutrons can be based either on the (D,T) accelerator reaction or from fission processes. The first option is limited in the source intensity (see Table 15.1) and requires long exposure time therefore. A new facility (NECTAR at FRM-2 [10]) is near to be operational with a good performance.

15.5 The Application of Neutron Imaging

As already demonstrated by the few examples in the text, the application of neutron is very versatile. Compared to common X-ray inspection, neutron imaging is applied favorably in such cases, when small amounts of organic material should be investigated within metallic covers. Both static and dynamic assemblies are nowadays able to be investigated in a good quality.

Any kind of moisture transport in structures (stone, wood, soil, plants, metals...) are the topic for research and industrial applications. In most cases, a detailed quantification of the moisture content is required and can be delivered by a dedicated analysis of the image data.

A prominent example is the moisture determination in electric fuel cells [11] where the water management plays an important role for the optimization of the performance of the cells. Solid forms of organics as adhesive, lacquer, and varnish are important fields for the application of neutron-imaging methods. An example is given in Fig. 15.12 for a glued part of an automotive car body.

Furthermore, the inspection of explosives from both military and civilian applications (space research, mining, tunnel construction,...) is very easily done with neutrons because of the high amount of organic compositions contained which would be more or less transparent for X-ray techniques.

The list cannot be complete because new technologies and new products are under permanent development. With the need for best possible reliability,

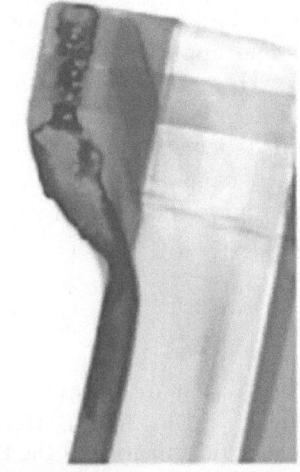

Fig. 15.12. Examples for practical applications of neutron imaging techniques: tomography of a car body where the adhesive connection is controlled (*dark area*); explosives like the cartridge of an air defence facility (*below*)

safety, and cost efficiency, the demand will increase for a fast and nondestructive inspection with sophisticated methods as neutron imaging can represent.

15.6 Future Trends and Visions

The future of neutron imaging as an important tool in science and technical applications depends much on the access to dedicated beam lines at the prominent neutron sources. Although clearly dominated by the neutron scattering community, a few new installations have recently been initiated at the centers ILL [8], FRM-2 [12], HMI [13], and PSI [14]. On the basis of this platform, it might become possible to support industry continuously with a reliable and powerful tool for a solution of many problems in nondestructive testing that would not be solved with other techniques in the same way. For the collaboration with research groups, completely different access conditions can be provided, similar to these common in neutron scattering.

15.7 Conclusions

It was shown that the field of neutron imaging provides promising aspects by the new class of digital detectors, by new imaging methods, and therefore several new approaches and applications. Based on the specific properties of neutrons in interaction with matter, neutron imaging methods will provide a value contribution in research and technology on the macroscopic scale (about $10\,\mu$m to 30 cm). It will depend importantly on the dialog and interaction between the responsible persons at the beam lines with partners in science and industrial companies how efficient and successful these methods will be progressively used in practice.

References

1. G. Bauer: J. Phys. IV France **9**, 91 (1999)
2. E. Lehmann, P. Vontobel: Applied Rad. Isotopes **61**, 567 (2004)
3. M. Estermann, J. Dubois: Investigation of the Properties of Amorphous Silicon Flat Panel Detectors suitable for Real-Time Neutron Imaging. In: *Proc. 7th World Conf. Neutron Radiography, Rome,* 2002
4. M. Balaskó et al.: Radiography inspection of helicopter rotor blades. In: *Conf. Proc. 3rd International Nondestructive Testing Conf.,* Crete, October 2003, pp. 309–313
5. P. Vontobel et al.: Neutrons for the study of adhesive connections. In: *Proc. 16th World Conference on Nondestructive Testing,* Montreal, 2004
6. N. Kardjilov: Dissertation, TU Munich, 2003, pp. 86–100
7. R. Hassanein et al.: Methods of scattering corrections for quantitative neutron radiography. In: *Proc. 5th Int. Topical Meeting on Neutron Radiography,* Munich, 2004

8. http://www.neutrograph.de/german/
9. N. Kardjilov: Dissertation, TU Munich, 2003, pp. 40–62
10. http://www.radiochemie.de/main/instr/nectar/nectar.html
11. D. Kramer et al.: An on-line study of fuel cell behavior by thermal neutrons. In: *Proc. 5th Int. Topical Meeting on Neutron Radiography*, Munich, 2004
12. http://www1.physik.tumuenchen.de/lehrstuehle/E21/e21_boeni.site/antares/web_new/
13. I. Manke et al.: The new cold neutron radiography and tomography facility at HMI and its industrial applications. In: *Proc. 5th Int. Topical Meeting on Neutron Radiography*, Munich, 2004
14. G. Kühne: CNR – The new beamline for cold neutron imaging at the Swiss spallation neutron source SINQ. In: *Proc. 5th Int. Topical Meeting on Neutron Radiography*, Munich, 2004

Index

Lecture Notes in Physics

For information about earlier volumes
please contact your bookseller or Springer
LNP Online archive: springerlink.com